Learning Linear Algebra
through DERIVE

Learning Linear Algebra through DERIVE

Brian Denton
School of Computing and Mathematical Sciences
Liverpool John Moores University

PRENTICE HALL

London   New York   Toronto   Sydney   Tokyo   Singapore
Madrid   Mexico City   Munich

First published 1995 by
Prentice Hall International (UK) Limited
Campus 400, Maylands Avenue
Hemel Hempstead
Hertfordshire HP2 7EZ
A division of
Simon & Schuster International Group

Printed at Redwood Books, Trowbridge, Wiltshire.

Library of Congress Cataloging-in-Publication Data

Available from the publisher

British Library Cataloguing in Publication Data

A catalogue record for this book is available from
the British Library

ISBN 0–13–122664–9

1   2   3   4   5   99   98   97   96   95

Contents

Contents

Contents

Series Preface

This series is designed to encourage the teaching and learning of various courses in mathematics using the symbolic algebra package DERIVE. Each text in the series will include the 'standard theory' appropriate to the subject, worked examples, exercises and *DERIVE Activities* so that students have a sound conceptual base on which to build. It is through the DERIVE activities that the learning of mathematics through investigation and discussion is encouraged. The aims of each DERIVE activity are to provide an investigation to introduce a new topic and to introduce the commands for using DERIVE as a manipulator in mathematical problem solving. Mathematics teaching and learning is changing as new technology becomes more readily available and these texts are written to support the new methods.

John Berry
Centre for Teaching Mathematics
University of Plymouth

Preface

This book came out of a two-semester course, each semester being approximately thirteen weeks' teaching. Examinations are set for the two weeks following the teaching. In the first semester the module is called simply 'Linear Algebra' and in the second semester the module is called 'Applied Linear Algebra'.

The modules are in the core for students reading mathematics, statistics and some computing degrees. The students in the main have been introduced to DERIVE in their first year before they take the linear algebra modules, so we can assume some knowledge of how to get around DERIVE (but not necessarily the matrices and determinants sections). Hence, in this book, I do not assume too much of the reader. In the early chapters rather full instructions are given as to what to do once DERIVE is loaded. As the book progresses we give less instruction except when a new idea is introduced or when perhaps a Utility File is needed.

I hope this book proves useful to a wide spectrum of students and they will feel able to study independently of a teacher for much of the time. What makes this book different from the plethora of linear algebra books is its integral use of DERIVE. The reason for using DERIVE is, as John Berry said in the preface to the first book in this series, that it is a package which combines power and 'user friendliness'. Most students are 'at home' with DERIVE after an hour or two in the laboratory. Also, DERIVE sits on a palm top computer and does not need the vast amount of memory other computer algebra packages require.

The book contains nine chapters. The contents list is comprehensive as is the index. The page numbers given in bold refer to the main appearance of the definition of a concept or idea. We have given an index of definitions separately for easy reference. Answers to most of the exercises and solutions to the consolidation exercises are given. If answers are not given it is thought that the problems are particularly straightforward.

I would welcome comments from teachers and students who use this book, especially about how it may be improved. Any constructive criticism will be welcomed.

Finally, can I thank a few people for their help and encouragement. Sharon Ward for her magnificent work in preparing the manuscript and John Berry for being a sympathetic series editor.

I dedicate this book to Pam, without whose understanding it might never have been written.

Brian H. Denton

School of Computing and Mathematical Sciences
Liverpool John Moores University
Byrom Street
Liverpool L3 3AF

E-mail: B.H.DENTON@LIVJM.AC.UK

1

Matrices

1.1 THE IDEA OF A MATRIX

A *matrix* (plural *matrices*) is an array of numbers or elements. Round parentheses or square brackets are used to hold the array. Capital letters are used to represent matrices. For example,

$$C = \begin{pmatrix} 1 & 2 \\ 3 & 4 \end{pmatrix} \qquad A = \begin{pmatrix} 3 & 1 & -2 & 4 \\ -1 & 0 & 3 & 0 \end{pmatrix}$$

are examples of matrices.

As DERIVE uses square brackets we will also use them throughout this book. The array is named by its *size* or *order*. For the matrices above, C is a 2×2 (two by two) matrix and A is a 2×4 (two by four) matrix. In general, we say that an $n \times m$ matrix has n rows and m columns. Matrices can be used to store information, for example in computer programming and statistics.

We will study matrices as objects, initially without reference to applications. The main aim of this chapter is to develop the algebra of matrices.

General Notation

An element of a matrix that stands in the ith row and the jth column is labelled a_{ij}. A matrix can then be written succinctly as $[a_{ij}]$, the size being indicated as $n \times m$ or 4×5, etc. Referring to matrices A and C above, $a_{12} = 1$ and $a_{23} = 3$, and $c_{11} = 1$ and $c_{22} = 4$.

A general 2×2 matrix would be

$$\begin{bmatrix} a_{11} & a_{12} \\ a_{21} & a_{22} \end{bmatrix}$$

So $[a_{ij}]$, given as a 2×2 matrix, would be written in full as above.

Another way of giving this information is to give the range of values for i and j, either separately or together in the case of a square matrix. If you were asked to write down the matrix $[a_{ij}]$ in full, where $0 \le i \le 2$ and $0 \le j \le 2$, the same matrix as given above would be the result. Alternatively, we could have given the one inequality $0 \le i, j \le 2$, to cover both ranges of values.

Exercise 1A

1.   State the order of the following matrices:

$$R = \begin{bmatrix} 1 & 0 \\ 4 & 2 \end{bmatrix} \qquad S = \begin{bmatrix} 1 \\ 0 \\ 1 \end{bmatrix} \qquad T = \begin{bmatrix} 7 & -1 & 3 & 2 \\ 4 & 1 & 0 & 6 \\ 5 & -2 & -1 & 0 \end{bmatrix}$$

2.   Let D be the matrix

$$\begin{bmatrix} 1 & 2 & 3 & 4 \\ 5 & 7 & 0 & 11 \end{bmatrix}$$

Write down each of the elements in D in ij notation. To get you started $d_{11} = 1$ and $d_{21} = 5$.

3.   Let E and F be the two matrices

$$\begin{bmatrix} 5 & 6 \\ 1 & -3 \end{bmatrix} \quad \text{and} \quad \begin{bmatrix} 0 & -1 \\ 5 & 2 \end{bmatrix}$$

respectively. Then $e_{11} + f_{11} = 5 + 0 = 5$.
Write down three similar expressions for the remaining elements.

4.   Write down, in full, the matrix $[a_{ij}]$ whose size is 3×4.

5.   Write down, in full, the matrix $[a_{ij}]$, where $0 \le i, j \le 3$.

1.2   ADDITION OF MATRICES

Two matrices can be added together if and only if they have the same size or order. The word we use to indicate whether we can add two matrices is *conformable*:

Matrices are *conformable for addition* if they have the same size.

We add two 2×3 matrices in the following way:

$$\begin{bmatrix} 1 & 2 & 3 \\ 2 & 5 & 7 \end{bmatrix} + \begin{bmatrix} 4 & 5 & 6 \\ 1 & 0 & 2 \end{bmatrix} = \begin{bmatrix} 1+4 & 2+5 & 3+6 \\ 2+1 & 5+0 & 7+2 \end{bmatrix} = \begin{bmatrix} 5 & 7 & 9 \\ 3 & 5 & 9 \end{bmatrix}$$

We observe that the size of the resulting matrix is the same as the size of the original two matrices. This will always be the case from the definition of addition.

Clearly, the two matrices A and C given on page 1 cannot be added because they are not the same size. The two matrices in Problem 3 of Exercise 1A can be added. The results of the additions asked for in this exercise would be the entries in the final matrix. Subtraction is defined in a similar way.

In general, we have the rule $[a_{ij}] \pm [b_{ij}] = [a_{ij} \pm b_{ij}]$, provided that the two matrices $[a_{ij}]$ and $[b_{ij}]$ are conformable. If we subtract the two matrices given above, the result would be as follows:

$$\begin{bmatrix} 1 & 2 & 3 \\ 2 & 5 & 7 \end{bmatrix} - \begin{bmatrix} 4 & 5 & 6 \\ 1 & 0 & 2 \end{bmatrix} = \begin{bmatrix} 1-4 & 2-5 & 3-6 \\ 2-1 & 5-0 & 7-2 \end{bmatrix} = \begin{bmatrix} -3 & -3 & -3 \\ 1 & 5 & 5 \end{bmatrix}$$

Example 1A

Which of the following matrices are conformable for addition?

$$P = \begin{bmatrix} 1 & 0 \\ 0 & 1 \\ -3 & 4 \end{bmatrix} \quad R = \begin{bmatrix} 1 & 0 \\ 4 & 2 \end{bmatrix} \quad S = \begin{bmatrix} 1 \\ 0 \\ 1 \end{bmatrix} \quad T = \begin{bmatrix} 4 & 7 \\ 3 & -1 \end{bmatrix} \quad U = \begin{bmatrix} -1 \\ 3 \\ -2 \end{bmatrix}$$

Solution

The order of the matrices is $P = 3 \times 2$, $R = 2 \times 2$, $S = 3 \times 1$, $T = 2 \times 2$, $U = 3 \times 1$. Hence R and T are conformable and S and U are conformable because each pair is of the same order.

DERIVE ACTIVITY 1A

Loading Matrices

(A)   Load DERIVE and press 'D' for the **Declare** option, then press 'M' for the
        Matrix option. The cursor will be flashing under the 3 in the rows position.
        DERIVE defaults to a 3×3 matrix so we must declare the size at the outset.
              To input the 2×4 matrix A in our first example, we proceed as follows.
        Type '2'. Now press the 'TAB' key to move to the columns field and type '4',
        followed by 'ENTER'. At this stage we are being asked what element we want
        for the position (1,1) or a_{11}. At the moment it holds 0, the default value.
        Enter '3' to obtain a_{11}. Next it will ask for a_{12}. We want this to be 1 so follow
        the sequence we used for a_{11} above to obtain a_{12}. Now continue and input a_{13}
        as –2, a_{14} as 4 and a_{21} as –1. As $a_{22} = 0$ in our required matrix, we simply
        press Enter, using the default value. Finish off the entries making $a_{23} = 3$ and
        $a_{24} = 0$. The screen shown in Figure 1.1 should now be displayed.

#1: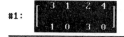

```
COMMAND: Author Build Calculus Declare Expand Factor Help Jump soLve Manage
         Options Plot Quit Remove Simplify Transfer Unremove moVe Window approX
Enter option
User                                    Free:100%                    Derive Algebra
```

Figure 1.1   The 2×4 matrix A

(B)   Enter the following matrix in the same way:

$$\begin{bmatrix} 0 & 1 & 2 & 3 \\ -1 & -2 & -3 & 2 \end{bmatrix}$$

You should have the display shown in Figure 1.2.

#1: $\begin{bmatrix} 3 & 1 & -2 & 4 \\ -1 & 0 & 3 & 0 \end{bmatrix}$

#2:

```
COMMAND: Author Build Calculus Declare Expand Factor Help Jump soLve Manage
         Options Plot Quit Remove Simplify Transfer Unremove moVe Window approX
Enter option
User                                    Free:100%                    Derive Algebra
```

Figure 1.2   DERIVE screen shows two matrices

Now enter matrix C from our first example, that is,

$$\begin{bmatrix} 1 & 2 \\ 3 & 4 \end{bmatrix}$$

Addition and Subtraction of Matrices

(C)   Whenever we need to use the matrix in Figure 1.1, we type '#1'. DERIVE
       reads what is on line 1. To obtain the sum of the two matrices we type
       #1 + #2. To do this **Author** #1 + #2. Figure 1.3 shows the DERIVE algebra
       window.

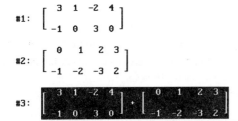

#1: $\begin{bmatrix} 3 & 1 & -2 & 4 \\ -1 & 0 & 3 & 0 \end{bmatrix}$

#2: $\begin{bmatrix} 0 & 1 & 2 & 3 \\ -1 & -2 & -3 & 2 \end{bmatrix}$

#3: $\begin{bmatrix} 3 & 1 & -2 & 4 \\ -1 & 0 & 3 & 0 \end{bmatrix} + \begin{bmatrix} 0 & 1 & 2 & 3 \\ -1 & -2 & -3 & 2 \end{bmatrix}$

```
COMMAND: Author Build Calculus Declare Expand Factor Help Jump soLve Manage
         Options Plot Quit Remove Simplify Transfer Unremove moVe Window approX
Enter option
User                                    Free:100%              Derive Algebra
```

Figure 1.3   The sum of the two matrices

For DERIVE to actually do the addition we now press 'S' for **Simplify**,
followed by 'ENTER'. Figure 1.4 should now be on the screen and the result of
the addition is on line 4.

#1: $\begin{bmatrix} 3 & 1 & -2 & 4 \\ -1 & 0 & 3 & 0 \end{bmatrix}$

#2: $\begin{bmatrix} 0 & 1 & 2 & 3 \\ -1 & -2 & -3 & 2 \end{bmatrix}$

#3: $\begin{bmatrix} 3 & 1 & -2 & 4 \\ -1 & 0 & 3 & 0 \end{bmatrix} + \begin{bmatrix} 0 & 1 & 2 & 3 \\ -1 & -2 & -3 & 2 \end{bmatrix}$

#4: $\begin{bmatrix} 3 & 2 & 0 & 7 \\ -2 & -2 & 0 & 2 \end{bmatrix}$

```
COMMAND: Author Build Calculus Declare Expand Factor Help Jump soLve Manage
         Options Plot Quit Remove Simplify Transfer Unremove moVe Window approX
Compute time: 0.0 seconds
Simp(#3)                                Free:100%              Derive Algebra
```

Figure 1.4   **Simplify** will show the value of the sum

Calling the matrix on line 1, A, and the matrix on line 2, B, work out the differences $A - B$ and $B - A$ using DERIVE. What happens when you ask DERIVE to add the matrix C in line 3 to either A or B? Try to explain the result you obtain on your screen.

(D)   Finally in this activity, we give a way of correcting an entry in a matrix if we have typed in a wrong number in a certain position. We suggest you start afresh with a clear screen, by removing all the five lines you have on your screen using the **Remove** command. Enter the matrix

$$\begin{bmatrix} 3 & 1 & -5 & 4 \\ -1 & 0 & 3 & 0 \end{bmatrix}$$

in line 1.

The error is in the entry a_{13}. It should be -2 if you consider the original 2×4 matrix in this activity. The matrix you wish to alter must be highlighted.

To alter the 5 to a 2 we proceed as follows.

Author F3 brings the matrix down into the author line in the following form:

$$[[3,1,-5,4], [-1,0,3,0]]$$

with the flashing cursor key at the end of the expression.

To move the cursor towards the -5 entry we hold down the control key 'Ctrl' and while holding it down hit the S key a number of times to move the cursor towards the left. Stop with the cursor on the comma after the -5 entry. Now press the delete key to remove the 5.

Later versions of DERIVE make use of the arrow keys to position the cursor. Before we proceed any further, make sure you are in **Insert** mode. To do this look at the space after the expression 'Free: 100%'. If it is empty, press the **Insert** key. The word Insert should now appear. (If Insert is already in the space, which is unlikely, then you do not need to do anything at this stage.) Now type '2' and you should have corrected the number entered in error. Finally, press Enter to display on line 2 the corrected matrix.

It is good working practice to remove the incorrect line using the **Remove** command.

1.3   SOME SPECIAL MATRICES

A matrix with the same number of rows as columns is called *a square matrix*. Hence we have 1×1, 2×2, 3×3 matrices, etc., all of which are square. These form a very useful subset of all matrices, as we will see later in applications. The elements a_{ii} are known as the *diagonal elements*.

Although a_{ii} is meaningful in rectangular arrays we keep the terminology of diagonal elements for square matrices only.

A matrix in which all the elements are zero is called *a zero matrix* or a *null matrix*. Thus there is an infinite number of zero matrices of different sizes. We have to be careful, therefore, to think of the correct size in applications. We use 0 to represent a zero matrix:

$$\begin{bmatrix} 0 & 0 \\ 0 & 0 \end{bmatrix} \qquad \text{is the } 2 \times 2 \text{ zero matrix}$$

$$\begin{bmatrix} 0 & 0 & 0 & 0 \\ 0 & 0 & 0 & 0 \end{bmatrix} \qquad \text{is the } 2 \times 4 \text{ zero matrix}$$

Using C and A from our first example, $C + 0$ and $A + 0$ will need zero matrices of size 2×2 and 2×4, respectively, as given above.

A square matrix in which all the diagonal elements are 1 and all other elements are 0 is called *a unit matrix* (usual notation I):

$$\begin{bmatrix} 1 & 0 \\ 0 & 1 \end{bmatrix} \qquad \text{is the } 2 \times 2 \text{ unit matrix}$$

$$\begin{bmatrix} 1 & 0 & 0 \\ 0 & 1 & 0 \\ 0 & 0 & 1 \end{bmatrix} \qquad \text{is the } 3 \times 3 \text{ unit matrix}$$

Using some of the notation we have introduced so far we could define a unit matrix as follows: a unit matrix is a square matrix in which

(i) $a_{ii} = 1$ for all i, and
(ii) $a_{ij} = 0$ for $i \neq j$.

1.4   EQUALITY OF MATRICES

Two matrices are equal if

(i)   they have the same size or order, and
(ii)   if the corresponding elements are equal.

Hence $A = B$ implies that A and B are the same size and $a_{ij} = b_{ij}$ for all i and j. If

$$A = \begin{bmatrix} 1 & 2 \\ 3 & 4 \end{bmatrix} \quad \text{and} \quad B = \begin{bmatrix} a & 2 \\ 3 & b \end{bmatrix}$$

then $A = B$ implies $a = 1$ and $b = 4$.

1.5   SOME MORE MATRICES

A square matrix is said to be *symmetric* if $a_{ij} = a_{ji}$ for all i, j:

$$\begin{bmatrix} 1 & 2 \\ 2 & 1 \end{bmatrix} \quad \text{and} \quad \begin{bmatrix} 1 & 2 & 3 \\ 2 & 4 & 9 \\ 3 & 9 & 7 \end{bmatrix}$$

are examples of symmetric matrices.

The *transpose* of a matrix A is obtained by interchanging the rows with the columns. Thus a 2×3 matrix becomes a 3×2 matrix and in general an $n \times m$ matrix becomes an $m \times n$ matrix when transposed. If A denotes the original matrix, then A' or A^T denotes the transpose of A. Thus if

$$A = \begin{bmatrix} 1 & 2 & 3 \\ 3 & 4 & 5 \end{bmatrix} \quad \text{then} \quad \begin{bmatrix} 1 & 3 \\ 2 & 4 \\ 3 & 5 \end{bmatrix} = A^T = A'$$

Notice that $(A^T)^T = A$ and if A is symmetric, then $A = A^T$, and conversely. The unit matrices are also examples of symmetric matrices.

DERIVE ACTIVITY 1B

In DERIVE, to obtain the transpose we use the back accent key (`), which is often located on the top row, left-hand side, of the keyboard. It should be noted that not all keyboards have such a key and then we have to find another way of entering the transpose of a matrix.

(A)   Input matrix A from our first example, that is,

$$\begin{bmatrix} 3 & 1 & -2 & 4 \\ -1 & 0 & 3 & 0 \end{bmatrix}$$

Then if the back accent key exists on your keyboard enter (#1) followed by the back accent, press S for **Simplify** and you should have the display shown in Figure 1.5.

#1: $\begin{bmatrix} 3 & 1 & -2 & 4 \\ -1 & 0 & 3 & 0 \end{bmatrix}$

#2: $\begin{bmatrix} 3 & 1 & -2 & 4 \\ -1 & 0 & 3 & 0 \end{bmatrix}$`

#3: $\begin{bmatrix} 3 & -1 \\ 1 & 0 \\ -2 & 3 \\ 4 & 0 \end{bmatrix}$

COMMAND: **Author** Build Calculus Declare Expand Factor Help Jump soLve Manage
          Options Plot Quit Remove Simplify Transfer Undelete moVe Window approX
Compute time: 0.1 seconds
Simp(#2)   Free:50%   Derive Algebra

Figure 1.5   The transpose of the original matrix using the back accent key (`)

The alternative is to use the **Build** command. Again start with matrix A in line 1. Now press 'B' then press Enter to obtain Figure 1.6, where line 1 and the + sign are highlighted.

#1: $\begin{bmatrix} 3 & 1 & -2 & 4 \\ -1 & 0 & 3 & 0 \end{bmatrix}$

BUILD: Operator: + - * / ^ . ▌ = Minus Recip Ln Exp Tan Sin Cos Atan ! % Done

Select operator
User   Free:100%   Derive Algebra

Figure 1.6   Using the **Build** command to find (`)

Now press the space bar until the highlight reaches the back accent symbol, `,
immediately before the equals sign, then press Enter. Press Enter again to see
the result on line 2.

Simplify and the transpose appears as #3, and Figure 1.5 should be
displayed.

(B)   Input other matrices and verify $(A^T)^T = A$ and $(A + B)^T = A^T + B^T$. Or in
DERIVE notation $(A')' = A$ and $(A + B)' = A' + B'$.

What is the condition on A for A and its transpose to be conformable
under addition?

1.6   THE ALGEBRA OF MATRICES

Consider again the 2×2 matrix

$$C = \begin{bmatrix} 1 & 2 \\ 3 & 4 \end{bmatrix}$$

If we now compute $C + C$ according to the rule given in Section 1.2 we obtain

$$\begin{bmatrix} 2 & 4 \\ 6 & 8 \end{bmatrix}$$

In algebra we have $C + C = 2C$. What does $2C$ mean in the context of matrices?
Does

$$2 \times \begin{bmatrix} 1 & 2 \\ 3 & 4 \end{bmatrix}$$

mean the same as

$$\begin{bmatrix} 2 & 4 \\ 6 & 8 \end{bmatrix}?$$

We define a matrix, A, multiplied by a scalar, for example a real or complex
number, as follows.

If $A = [a_{ij}]$ and λ is any scalar we define $\lambda A = [\lambda a_{ij}]$. In simple language this
means that every element in the matrix is multiplied by the scalar λ. In the example
given above we see that $2C$ obeys this definition.

Example 1B

For the following matrices A and B work out, $-3A$, $\frac{1}{2}B$, $-3A + \frac{1}{2}B$:

$$A = \begin{bmatrix} -1 & 0 & 0 \\ 1 & 2 & 4 \\ 3 & -3 & 2 \end{bmatrix} \qquad B = \begin{bmatrix} 2 & -1 & 6 \\ -2 & 0 & 3 \\ 4 & 1 & -8 \end{bmatrix}$$

Solution

To evaluate $-3A$ we multiply each element of A by -3 to give

$$-3A = \begin{bmatrix} 3 & 0 & 0 \\ -3 & -6 & -12 \\ -9 & 9 & -6 \end{bmatrix}$$

To evaluate $\frac{1}{2}B$ we multiply each element of B by $\frac{1}{2}$ to give

$$\tfrac{1}{2}B = \begin{bmatrix} 1 & -\tfrac{1}{2} & 3 \\ -1 & 0 & \tfrac{3}{2} \\ 2 & \tfrac{1}{2} & -4 \end{bmatrix}$$

Since $-3A$ and $\frac{1}{2}B$ are conformable we can add them term-by-term:

$$-3A + \tfrac{1}{2}B = \begin{bmatrix} 4 & -\tfrac{1}{2} & 3 \\ -4 & -6 & -10\tfrac{1}{2} \\ -7 & 9\tfrac{1}{2} & -10 \end{bmatrix}$$

Exercise 1B

1.   If $A = \begin{bmatrix} -1 & -3 \\ -2 & 4 \end{bmatrix}$ then $5A = \begin{bmatrix} -5 & -15 \\ -10 & 20 \end{bmatrix}$. Write down $-2A$ and $A/2$.

2.   If we write $-A$ for $(-1)A$ we have $A + -A = 0$ for any matrix A, where 0 is the zero matrix having the same size as A. Write down $-A$ using the matrix in 1 and verify the statement $A + -A = 0$.

3.   Can we always find a matrix B such that $A + B = 0$, where A, B and 0 all have
      the same size?
4.   For the following pair of matrices calculate $2P - 3R$ and $0.7P + 0.2R$:

$$P = \begin{bmatrix} 1 & -1 \\ 3 & 2 \end{bmatrix} \qquad R = \begin{bmatrix} 0 & 6 \\ 4 & -5 \end{bmatrix}$$

Use DERIVE to check all your answers.

1.7   MULTIPLICATION OF MATRICES

To motivate the way in which two matrices can be multiplied together consider the
following matrices in a shopping application, using the language of economics.
      Consider the situation of two people, Mrs A and Mr B, doing some shopping
over three weeks. In particular we pick just two of the commodities in their shopping
baskets, butter and cheese:

$$\begin{array}{c} \\ \text{Mrs } A \\ \text{Mr } B \end{array} \begin{array}{cc} \text{Butter} & \text{Cheese} \\ \begin{bmatrix} 2 & 3 \\ 4 & 3 \end{bmatrix} \end{array} = C$$

We call this the commodity matrix for the two customers. The entries would mean,
for example, that Mrs A buys two units of butter each week and three units of cheese,
and Mr B buys four units of butter and three units of cheese each week.
      The following is the price matrix for the three weeks:

$$\begin{array}{c} \\ \text{Butter} \\ \text{Cheese} \end{array} \begin{array}{ccc} \text{Wk 1} & \text{Wk 2} & \text{Wk 3} \\ \begin{bmatrix} 5 & 4 & 3 \\ 2 & 3 & 2 \end{bmatrix} \end{array} = P$$

The first column entries for butter and cheese would be the price per unit for the first
week; similarly for the second and third columns. With this information we work out
the cost to the two customers for each week and display the total cost in a third matrix
given below:

Mrs A Mr B
in Wk 1 spends $2 \times 5 + 3 \times 2 = 16$ in Wk 1 spends $4 \times 5 + 3 \times 2 = 26$
in Wk 2 $2 \times 4 + 3 \times 3 = 17$ in Wk 2 $4 \times 4 + 3 \times 3 = 25$
in Wk 3 $2 \times 3 + 3 \times 2 = 12$ in Wk 3 $4 \times 3 + 3 \times 2 = 18$

The total cost matrix will then be as follows:

$$
\begin{array}{ccc}
\text{Wk 1} & \text{Wk 2} & \text{Wk 3}
\end{array}
$$

$$
\begin{array}{c}
\text{Mrs } A \\
\text{Mr } B
\end{array}
\begin{bmatrix}
16 & 17 & 12 \\
26 & 25 & 18
\end{bmatrix}
$$

In summary we have

$$
\begin{bmatrix} 2 & 3 \\ 4 & 3 \end{bmatrix} \cdot \begin{bmatrix} 5 & 4 & 3 \\ 2 & 3 & 2 \end{bmatrix} = \begin{bmatrix} 16 & 17 & 12 \\ 26 & 25 & 18 \end{bmatrix}
$$

Notice that a dot is used for multiplication because in DERIVE no other symbol is understood. (In later versions of DERIVE, juxtaposition will be understood and the dot is then not necessary.)

This example can now be used to analyze the process of multiplication of two matrices.

Observations

(i)   We observe that we will need as many rows in the price matrix (P) as we have columns in the commodity matrix (C); otherwise we cannot work out the final costs. (Each commodity needs a price.)

(ii)   The number of rows in C is immaterial; we can have a hundred customers if we wish and then C would have size 100×2.

(iii)   The number of columns in P is immaterial; we could go on for a year and P would then have size 2×52.

(iv)   The total cost matrix, CP, must be a 2×3 matrix (two customers for three weeks' prices).

(v)   The size of the final matrix, CP, is obtained by considering the sizes of C and P, as follows:

$$
C\,(2 \times 2) \times P\,(2 \times 3) = CP\,(2 \times 3)
$$

(i)

Size of product

If C had size 100×2 and P had size 2×52 then CP would have size 100×52.

Notice that we write CP for $C \cdot P$ as in ordinary algebra (compare this with xy).

Generally

If our commodity matrix, C, is $n \times m$ and our price matrix, P, is $m \times r$, then our total cost matrix, CP, is $n \times r$. Hence in our definition we need to have the fact that the columns of the first matrix must be equal in number to the rows in the second matrix.

> Two matrices are *conformable for multiplication* if the number of rows of one of them is equal to the number of columns of the other.

Exercise 1C

For each of the five pairings of matrices A and B below, state which are and which are not conformable for multiplication.

If A and B are conformable give the size of AB and /or BA. The first two have been done for you.

	(i)	(ii)	(iii)	(iv)	(v)
Matrix A	2×2	2×3	3×2	3×3	4×5
Matrix B	2×2	2×2	2×3	3×4	2×3

(i) Here both AB and BA can be worked out and the size of both will be 2×2.

(ii) Here AB cannot be worked out since the number of columns of A is 3, yet the number of rows of B is 2. However, BA is possible. Its size will be 2×3.

Important Observation

It follows from the above that AB cannot equal BA in general because one of these products might exist and the other one might not. Even if they both exist AB does not equal BA in general. This is the first example of matrices **not** obeying the 'usual' laws of algebra.

Notation

For $A \cdot A$ we write A^2 and thus A^3 means $A \cdot A \cdot A$, and similarly for A^n , where n is a positive integer. This means that the matrix A must be square for if it was $n \times r$ then A^2 would be $[n \times r] \cdot [n \times r]$ and for conformability $r = n$.

DERIVE ACTIVITY 1C

First do the problems below by hand and then check your answers using DERIVE. To input a product of two matrices which are on line 1 and line 2 **Author #1.#2.**
Given

$$A = \begin{bmatrix} 1 & -1 & 2 \\ 0 & 3 & 4 \end{bmatrix} \qquad B = \begin{bmatrix} 4 & 0 & -3 \\ -1 & -2 & 3 \end{bmatrix}$$

$$D = \begin{bmatrix} 2 \\ -1 \\ 3 \end{bmatrix} \qquad C = \begin{bmatrix} 2 & -3 & 0 & 1 \\ 5 & -1 & -1 & 2 \\ -1 & 0 & 0 & 3 \end{bmatrix}$$

work out the following expressions where possible:

(i) $A + B$ (ii) $C + D$ (iii) $3B$ (iv) AB
(v) BC (vi) CD (vii) D^2 (viii) AD
(ix) $D^T D$ (x) DD^T (xi) $C^T D^T$ (xii) $B - A$
(xiii) ABC

1.8   INVERSE OF A MATRIX UNDER ADDITION AND UNDER MULTIPLICATION

In Section 1.6, we had the equation $A + -A = 0$. We say that A and $-A$ are *inverses* of each other *under addition*. Alternatively, we say that $-A$ is the *additive inverse* of A, and conversely. Compare this notation to equations of the form $3 + {}^-3 = 0$. We say that 3 and $^-3$ are additive inverses of each other and their sum is zero by definition.
    Now we introduce *inverses under multiplication*. We require a matrix B such that $AB = BA = I$, the identity matrix of the correct size. If such a matrix exists, we write $B = A^{-1}$. Compare this notation with $2 \times 2^{-1} = 1$ and, in general, $xx^{-1} = 1$. It follows from the above that A must be a square matrix and A^{-1} is the same size as A. Not every matrix necessarily has a multiplicative inverse. For example, what is the multiplicative inverse of the zero matrix? Can we find a matrix A such that $A \times 0 = I$? In other words can we find the entries a_{ij} in the first matrix below which satisfy the equation?

$$\begin{bmatrix} a_{11} & a_{12} \\ a_{21} & a_{22} \end{bmatrix} \cdot \begin{bmatrix} 0 & 0 \\ 0 & 0 \end{bmatrix} = \begin{bmatrix} 1 & 0 \\ 0 & 1 \end{bmatrix}$$

The answer is no to these last two questions since $a_{11} \times 0 + a_{12} \times 0$ should equal 1 but in fact it equals 0.

Here is an example of a matrix A, which has an inverse. Consider the product of the two matrices

$$\begin{bmatrix} 2 & 4 \\ 1 & 3 \end{bmatrix} \text{ and } \begin{bmatrix} 1.5 & -2 \\ -0.5 & 1 \end{bmatrix}$$

We have

$$\begin{bmatrix} 2 & 4 \\ 1 & 3 \end{bmatrix} \cdot \begin{bmatrix} 1.5 & -2 \\ -0.5 & 1 \end{bmatrix} = \begin{bmatrix} 1 & 0 \\ 0 & 1 \end{bmatrix}$$

Hence if

$$A = \begin{bmatrix} 2 & 4 \\ 1 & 3 \end{bmatrix}$$

then from the above equation the inverse of A is given by

$$A^{-1} = \begin{bmatrix} 1.5 & -2 \\ -0.5 & 1 \end{bmatrix} = \frac{1}{2}\begin{bmatrix} 3 & -4 \\ -1 & 2 \end{bmatrix}$$

Summary

A square matrix A has a multiplicative inverse, A^{-1}, if and only if the following equations hold

$AA^{-1} = I = A^{-1}A.$

Square matrices which have multiplicative inverses are said to be *invertible* or *non-singular*. Those which do not have multiplicative inverses are said to be *singular*.

DERIVE ACTIVITY 1D

Use DERIVE to help you answer this question.

If

$$A = \begin{bmatrix} 1 & 2 \\ 3 & 4 \end{bmatrix} \text{ and } B = \begin{bmatrix} 1 & 2 \\ 2 & 4 \end{bmatrix}$$

do A^{-1} and B^{-1} exist?

To input the inverse A^{-1}, **Author** #1^(−1), assuming the matrix A is in line 1 of the DERIVE Algebra window.
        Investigate other 2×2 matrices and separate out those which have got multiplicative inverses and those which have not. Can you find anything in common between the matrices which do not have inverses? Can you find anything in common between the matrices which do have inverses?

Consolidation Exercise

To understand the mathematics it is advisable to attempt these exercises by hand and verify the results using DERIVE where applicable. Remember, in this exercise, A^T means the transpose of A.

1.   If $A = \begin{bmatrix} 1 \\ 2 \end{bmatrix}$ (a column matrix) and $B = \begin{bmatrix} 3 & 4 \end{bmatrix}$ (a row matrix) where possible

      work out:

(i)	$A + B$	(ii)	$A \cdot B$	(iii)	$B \cdot A$
(iv)	$2B - 3A$	(v)	$(A \cdot B)^T$	(vi)	$B^T \cdot A^T$

2.   If $C = \begin{bmatrix} -2 & -1 & 5 \\ 3 & 0 & 1 \\ 4 & -3 & 0 \end{bmatrix}$ and $D = \begin{bmatrix} -1 & 2 & -3 \end{bmatrix}$ where possible work out

(i)	$C \cdot D$	(ii)	C^T	(iii)	D^T
(iv)	$(D \cdot C)^T$	(v)	$C^T \cdot D^T$	(vi)	C^2

3.   Make a conjecture from your answers to 1(v) and 1(vi) and also from 2(iv) and 2(v) above. Try to generalize your result.

4.   If $E = \begin{bmatrix} -2 & 4 \\ 6 & -5 \end{bmatrix}$ and $\lambda = 3$ work out

(i) λE (ii) $E - \lambda I$ (iii) $5\lambda E^2$
(iv) $5(\lambda E)^2$ (v) $5\lambda^2 E^2$

Comment on your results for (iii), (iv) and (v).

5.   Let A be a square matrix with the property that

$$A^{m+1} = 0$$

where 0 is the appropriate zero matrix and m is some positive integer.
        Show that the matrix $(I - A)$ has an inverse, where I is the appropriate identity matrix.
        Write down an expression for $(I - A)^{-1}$.
        (Hints. This question can be answered by considering matrix algebra only. Do not write down a particular matrix for A because the chances are that the matrix A^{m+1} will not be zero for any value of m.
        Consider factorizing matrix expressions as, for example, $A^2 - B^2$ and $A^3 - B^3$ and then generalizing to $A^n - B^n$ with an appropriate substitution for B.
        Remember that solutions are given at the back of the book, but try to solve such questions as the one above over a period of time rather than in one sitting! Only when you have struggled with it should you approach the solution for a further hint.
        This question is one where DERIVE will not be that helpful!

2

Determinants

2.1   LINEAR EQUATIONS IN TWO UNKNOWNS

In this section we are going to study two linear equations in two unknowns. We will consider the problem in general terms. The unknowns in what follows are x and y and the a_{ij}s are assumed to be given real numbers. We wish to solve the equations 'simultaneously', that is, find a value for x and a value for y which satisfies both the equations.

Consider the following two equations:

$$a_{11}x + a_{12}y = k_1 \tag{2.1}$$
$$a_{21}x + a_{22}y = k_2 \tag{2.2}$$

This system of equations can be written in matrix form as follows:

$$\begin{bmatrix} a_{11} & a_{12} \\ a_{21} & a_{22} \end{bmatrix} \cdot \begin{bmatrix} x \\ y \end{bmatrix} = \begin{bmatrix} k_1 \\ k_2 \end{bmatrix}$$

Check this claim by working the product out by hand using the results from Chapter 1. Do not use DERIVE at this stage as a_{ij}, though possible to input, becomes a little clumsy, visually.

The matrix containing the a_{ij} elements is called the *matrix of coefficients*. We can eliminate x from the two equations by multiplying Equation 2.1 by a_{21} and by multiplying Equation 2.2 by a_{11} to obtain

$$a_{21}a_{11}x + a_{21}a_{12}y = a_{21}k_1 \tag{2.3}$$
$$a_{11}a_{21}x + a_{11}a_{22}y = a_{11}k_2 \tag{2.4}$$

and now subtract Equation 2.3 from Equation 2.4 to obtain

$$y(a_{11}a_{22} - a_{21}a_{12}) = (a_{11}k_2 - a_{21}k_1)$$

Hence,

$$y = (a_{11}k_2 - a_{21}k_1)/(a_{11}a_{22} - a_{21}a_{12})$$

In a similar way we can eliminate y from the two equations by multiplying Equation 2.1 by a_{22} and Equation 2.2 by a_{12} and then subtracting the two resulting equations to obtain

$$x = (a_{22}k_1 - a_{12}k_2)/(a_{11}a_{22} - a_{21}a_{12})$$

It can be seen that x and y would not exist if the denominator, which is the same in both expressions for x and y, was zero. We call the expression $(a_{11}a_{22} - a_{21}a_{12})$ the *determinant* of the matrix of coefficients $\begin{bmatrix} a_{11} & a_{12} \\ a_{21} & a_{22} \end{bmatrix}$ since it determines whether or not a unique solution exists for the system of equations.

Notation

If $A = \begin{bmatrix} a_{11} & a_{12} \\ a_{21} & a_{22} \end{bmatrix}$ then $\mathrm{DET}(A) = \begin{vmatrix} a_{11} & a_{12} \\ a_{21} & a_{22} \end{vmatrix} = a_{11}a_{22} - a_{21}a_{12}$.

Now that we have introduced determinants we can rewrite the unique solution to the two equations as follows:

$$x = \frac{\begin{vmatrix} k_1 & a_{12} \\ k_2 & a_{22} \end{vmatrix}}{\mathrm{DET}(A)}, \quad y = \frac{\begin{vmatrix} a_{11} & k_1 \\ a_{21} & k_2 \end{vmatrix}}{\mathrm{DET}(A)} \tag{2.5}$$

Note that a unique solution exists if and only if $\mathrm{DET}(A) \neq 0$. This gives us a quick way of checking whether or not two equations in two unknowns have a unique solution. Furthermore, the value of a determinant is a scalar, for example a real number.

Exercise 2A

Consider the following linear equations:

$$2x + 3y = 5$$
$$3x + 5y = 6$$

Write down the matrix of coefficients and work out its determinant. Does the system of equations have a unique solution? If your answer is 'yes', use the equations in (2.1) above to work out the unique values for x and y but if your answer is 'no' check the matrix of coefficients and its determinant!

DERIVE ACTIVITY 2A

(A)   Input the matrix of coefficients into line 1. DERIVE has as one of its functions DET(A), which gives the value of the determinant for the square matrix A. From our definition it is clear that the matrix A must be square. Use this function to evaluate the determinant in the exercise above using **Author** DET(#1) and **Simplify**. You should get the screen in Figure 2.1.

$$\#1: \quad \begin{bmatrix} 2 & 3 \\ 3 & 5 \end{bmatrix}$$

$$\#2: \quad \text{DET} \begin{bmatrix} 2 & 3 \\ 3 & 5 \end{bmatrix}$$

#3: []

```
COMMAND: Author Build Calculus Declare Expand Factor Help Jump soLve Manage
         Options Plot Quit Remove Simplify Transfer Unremove moVe Window approX
Compute time: 0.0 seconds
Simp(#2)                              Free:100%              Derive Algebra
```

Figure 2.1 **Author** DET(#1) gives the determinant

(B)   Input other 2×2 matrices and work out their determinants. Again, you are advised to work out the values for yourself first and check them using DERIVE.

Input a 2×2 matrix whose determinant is zero. (You may have already done so!) Write down two equations which would have your matrix as its matrix of coefficients. What can you say about the 'solution' to your equations?

2.2   LINEAR EQUATIONS IN THREE OR MORE UNKNOWNS

The idea introduced in Section 2.1 extends to the 3×3 case and on to the $n \times n$ case. Be warned that the algebra can be a little formidable and the subscripts on the 'a' terms must be recorded accurately, but the effort is well worth while!

     If we start with three equations with three unknowns we can eliminate y and z, for example, as before and gain an expression for x. Similarly, we could eliminate x and z to obtain an expression for y and finally eliminate x and y to obtain an expression for z.

     It should not be too much of a surprise to find that the expressions for x, y and z all have the same denominator and that we define this common expression to be the determinant of the 3×3 matrix of coefficients. Again a unique solution exists if and only if this determinant is non-zero. Consider, then, the three equations in the three unknowns x, y and z:

$$a_{11}x + a_{12}y + a_{13}z = k_1 \tag{2.6}$$
$$a_{21}x + a_{22}y + a_{23}z = k_2 \tag{2.7}$$
$$a_{31}x + a_{32}y + a_{33}z = k_3 \tag{2.8}$$

In matrix form this system can be written

$$\begin{bmatrix} a_{11} & a_{12} & a_{13} \\ a_{21} & a_{22} & a_{23} \\ a_{31} & a_{32} & a_{33} \end{bmatrix} \cdot \begin{bmatrix} x \\ y \\ z \end{bmatrix} = \begin{bmatrix} k_1 \\ k_2 \\ k_3 \end{bmatrix}$$

Again be sure to check that our claim is true.

     Now we will eliminate z from Equations (2.6) and (2.7), and then from Equations (2.7) and (2.8) to obtain two equations in x and y. Multiply Equation (2.6) by a_{23} and Equation (2.7) by a_{13} to obtain

$$a_{23}a_{11}x + a_{23}a_{12}y + a_{23}a_{13}z = a_{23}k_1$$
$$a_{13}a_{21}x + a_{13}a_{22}y + a_{13}a_{23}z = a_{13}k_2$$

Now subtract these two equations to eliminate the z term: hence,

$$(a_{23}a_{11} - a_{13}a_{21})x + (a_{23}a_{12} - a_{13}a_{22})y = a_{23}k_1 - a_{13}k_2$$

We can write this equation using determinants as follows:

$$x \begin{vmatrix} a_{11} & a_{13} \\ a_{21} & a_{23} \end{vmatrix} + y \begin{vmatrix} a_{12} & a_{13} \\ a_{22} & a_{23} \end{vmatrix} = \begin{vmatrix} k_1 & a_{13} \\ k_2 & a_{23} \end{vmatrix}$$

For convenience let us write this equation as

$$xD_1 + yD_2 = D_3 \tag{2.9}$$

Note down any common patterns that you observe between the three determinants, D_1, D_2 and D_3. Your patterns, or observations, should help you to remember this equation by simply looking at the original three equations. If you have observed enough features, then you should be able to write down a similar expression if you eliminate the z variable from Equations (2.7) and (2.8). Check your expression with the following:

$$x \begin{vmatrix} a_{21} & a_{23} \\ a_{31} & a_{33} \end{vmatrix} + y \begin{vmatrix} a_{22} & a_{23} \\ a_{32} & a_{33} \end{vmatrix} = \begin{vmatrix} k_2 & a_{23} \\ k_3 & a_{33} \end{vmatrix}$$

Again for convenience let us write this equation as

$$xD_4 + yD_5 = D_6 \tag{2.10}$$

Finally, we eliminate y from Equations (2.9) and (2.10) to obtain an expression for x. We now use the result obtained in the 2×2 case earlier to write down

$$x = \frac{D_3 D_5 - D_2 D_6}{D_1 D_5 - D_2 D_4}$$

Now carefully evaluate the denominator back to terms in a_{ij} to obtain the expression given below:

$$(a_{11}a_{23} - a_{21}a_{13})(a_{22}a_{33} - a_{32}a_{23}) - (a_{12}a_{23} - a_{13}a_{22})(a_{21}a_{33} - a_{23}a_{31})$$

$$= a_{11}a_{23}a_{22}a_{33} - a_{11}a_{23}^2 a_{32} - a_{21}a_{13}a_{22}a_{33} + a_{21}a_{13}a_{32}a_{23}$$

$$- a_{12}a_{23}a_{21}a_{33} + a_{12}a_{23}^2 a_{31} + a_{13}a_{22}a_{21}a_{33} - a_{13}a_{22}a_{23}a_{31}$$

Observe that the third and seventh terms cancel, leaving us with six terms, each with a common factor of a_{23}. Thus we can simplify and factorize to obtain,

$$a_{23}(a_{11}a_{22}a_{33} - a_{11}a_{23}a_{32} + a_{21}a_{13}a_{32} - a_{12}a_{21}a_{33} + a_{12}a_{23}a_{31} - a_{13}a_{22}a_{31})$$

Now leave this expression and let us turn our attention to the numerator in the x expression.

We evaluate the numerator back to terms in a_{ij} and k to obtain

$$(k_1a_{23} - k_2a_{13})(a_{22}a_{33} - a_{23}a_{32}) - (a_{12}a_{23} - a_{13}a_{22})(k_2a_{33} - k_3a_{23})$$

$$= k_1a_{23}a_{22}a_{33} - k_1a_{23}^2a_{32} - k_2a_{13}a_{22}a_{33} + k_2a_{13}a_{23}a_{32}$$

$$-k_2a_{12}a_{23}a_{33} + k_3a_{12}a_{23}^2 + k_2a_{33}a_{13}a_{22} - k_3a_{23}a_{13}a_{22}$$

Again observe that two terms cancel to leave six terms with a common factor of a_{23}. Thus the numerator simplifies and factorizes to

$$a_{23}(k_1a_{22}a_{33} - k_1a_{23}a_{32} + k_2a_{13}a_{32} - k_2a_{12}a_{33} + k_3a_{12}a_{23} - k_3a_{13}a_{22})$$

Assuming without loss of generality that a_{23} is not zero we can cancel this factor from the numerator and denominator of the x value, leaving us with a rather large expression for x (but do not give in...).

We now have the following quotient for x:

$$\frac{(k_1a_{22}a_{33} - k_1a_{23}a_{32} + k_2a_{13}a_{32} - k_2a_{12}a_{33} + k_3a_{12}a_{23} - k_3a_{13}a_{22})}{(a_{11}a_{22}a_{33} - a_{11}a_{23}a_{32} + a_{21}a_{13}a_{32} - a_{12}a_{21}a_{33} + a_{12}a_{23}a_{31} - a_{13}a_{22}a_{31})}$$

We will soon illustrate a simple way to remember this expression.

Exercise 2B

We have just eliminated z from Equations (2.6) and (2.7) and then from Equations (2.7) and (2.8) to obtain the two equations in x and y below:

$$xD_1 + yD_2 = D_3$$
$$xD_4 + yD_5 = D_6$$

Then we obtained an expression for x using previous results. Using the two equations above and a result from our 2×2 work earlier, we obtain a similar expression for y.

Finally, using the expressions for x and y or using a different elimination obtain a value for z which looks similar to the ones for x and y. (This is a messy exercise but well worth doing once!)

Show from your results that the denominators of x, y and z are equal and that the common expression is given by the following six terms in some order:

$$a_{11}a_{22}a_{33} - a_{11}a_{23}a_{32} + a_{13}a_{21}a_{32} - a_{21}a_{33}a_{12} + a_{31}a_{23}a_{12} - a_{31}a_{13}a_{22}$$

This expression is the determinant of the matrix of coefficients of the Equations (2.6), (2.7) and (2.8). This means that if

$$A = \begin{bmatrix} a_{11} & a_{12} & a_{13} \\ a_{21} & a_{22} & a_{23} \\ a_{31} & a_{32} & a_{33} \end{bmatrix}$$

is the matrix of coefficients, then the determinant of A, DET(A), which is written as

$$\begin{vmatrix} a_{11} & a_{12} & a_{13} \\ a_{21} & a_{22} & a_{23} \\ a_{31} & a_{32} & a_{33} \end{vmatrix}$$

is evaluated as

$$a_{11}a_{22}a_{33} - a_{11}a_{23}a_{32} - a_{21}a_{33}a_{12} + a_{31}a_{23}a_{12} + a_{13}a_{21}a_{32} - a_{31}a_{13}a_{22}$$

At this stage we may ask how do we remember such an expression in order to evaluate a 3×3 determinant? We did promise that it would be simple!

The six terms can be collected together in pairs and factorized in a number of different ways. We give below two of these possibilities.

Case 1

We use the elements of the first row of the matrix of coefficients, a_{11}, a_{12} and a_{13} as factors and gather the six terms into three as follows:

$$a_{11}(a_{22}a_{33} - a_{32}a_{23}) - a_{12}(a_{21}a_{33} - a_{31}a_{23}) + a_{13}(a_{21}a_{32} - a_{31}a_{22})$$

Notice that this can be written in determinant form as

$$a_{11}\begin{vmatrix} a_{22} & a_{23} \\ a_{32} & a_{33} \end{vmatrix} - a_{12}\begin{vmatrix} a_{21} & a_{23} \\ a_{31} & a_{33} \end{vmatrix} + a_{13}\begin{vmatrix} a_{21} & a_{22} \\ a_{31} & a_{32} \end{vmatrix}$$

and this expression, as we shall soon see, can be obtained directly from

$$\begin{vmatrix} a_{11} & a_{12} & a_{13} \\ a_{21} & a_{22} & a_{23} \\ a_{31} & a_{32} & a_{33} \end{vmatrix} = \text{DET}(A)$$

A second case is given below so that patterns should start to emerge.

Case 2

In this case we use the elements of the third column, a_{13}, a_{23} and a_{33} as factors and gather the six terms into three as follows:

$$a_{13}(a_{21}a_{32} - a_{31}a_{22}) - a_{23}(a_{11}a_{32} - a_{31}a_{12}) + a_{33}(a_{11}a_{22} - a_{21}a_{12})$$

This can be written in determinant form as

$$a_{13}\begin{vmatrix} a_{21} & a_{22} \\ a_{31} & a_{32} \end{vmatrix} - a_{23}\begin{vmatrix} a_{11} & a_{12} \\ a_{31} & a_{32} \end{vmatrix} + a_{33}\begin{vmatrix} a_{11} & a_{12} \\ a_{21} & a_{22} \end{vmatrix}$$

and as in Case 1 this expression can be written, as we shall see, from

$$\begin{vmatrix} a_{11} & a_{12} & a_{13} \\ a_{21} & a_{22} & a_{23} \\ a_{31} & a_{32} & a_{33} \end{vmatrix} = \text{DET}(A)$$

In Case 1 we say that we have expanded the determinant by the first row and in Case 2 we have expanded the determinant by the third column. We could have used any of the three rows or three columns to expand DET(A). If we took time to expand by the other two rows and two columns we would be given strong evidence for remembering what to do to obtain the value of the determinant.

Perhaps you have observed a pattern from the two cases above. The general pattern will be stated later in this chapter after we have introduced a few more ideas.

If you completed Exercise 2B you should be able to see that if a unique solution to our three equations exists, then it can be given in the form shown below (compare this with the 2×2 case):

$$x = \frac{\begin{vmatrix} k_1 & a_{12} & a_{13} \\ k_2 & a_{22} & a_{23} \\ k_3 & a_{32} & a_{33} \end{vmatrix}}{\mathrm{DET}(A)}, \qquad y = \frac{\begin{vmatrix} a_{11} & k_1 & a_{13} \\ a_{21} & k_2 & a_{23} \\ a_{31} & k_3 & a_{33} \end{vmatrix}}{\mathrm{DET}(A)}, \qquad z = \frac{\begin{vmatrix} a_{11} & a_{12} & k_1 \\ a_{21} & a_{22} & k_2 \\ a_{31} & a_{32} & k_3 \end{vmatrix}}{\mathrm{DET}(A)}$$

This rule is known as Cramer's rule and we have developed it to represent the unique solution case of three equations in three unknowns provided that $\mathrm{DET}(A) \neq 0$. It extends to n equations in n unknowns and, provided that $\mathrm{DET}(A) \neq 0$, gives the unique solution to the n equations.

Exercise 2C

1.   Solve the following system of equations:

$$2x + 3y - z = 1$$
$$3x + 5y + 2z = 8$$
$$x - 2y - 3z = -1$$

Hints for solution

The matrix of coefficients is

$$\begin{bmatrix} 2 & 3 & -1 \\ 3 & 5 & 2 \\ 1 & -2 & -3 \end{bmatrix} = A$$

First evaluate $\mathrm{DET}(A)$ by hand using any of the cases discussed above and check your result using DERIVE. You should find that $\mathrm{DET}(A) = 22$.

Hence Cramer's rule can be used. Using this rule work out the 'numerator determinants' for x, y and z; these should give 66, –22 and 44 respectively. Hence, $x = 3$, $y = -1$ and $z = 2$ is the unique solution.

2.   Evaluate the following 3×3 determinants using the two cases described above:

$$\begin{vmatrix} 1 & 0 & 0 \\ 2 & 3 & 6 \\ -1 & 4 & -1 \end{vmatrix} \qquad \begin{vmatrix} 0 & 1 & 2 \\ -1 & 2 & 0 \\ 3 & 4 & -2 \end{vmatrix} \qquad \begin{vmatrix} 5 & -1 & 3 \\ 2 & -2 & 1 \\ 4 & 0 & 7 \end{vmatrix}$$

Check your answers using DERIVE.

2.3   COFACTORS AND MINORS

Every element in a determinant has a *cofactor* and a *minor* associated with it. We obtain the minor of a_{ij} by deleting the row and the column in which a_{ij} stands and then working out the determinant that then remains. Hence minors of a 3×3 determinant are 2×2 determinants.

Example 2A

For the following determinant write down the minors for the elements a_{11}, a_{23} and a_{44}:

$$\begin{vmatrix} \textcircled{1} & 7 & 0 & -3 \\ 4 & 6 & \textcircled{-2} & 5 \\ 0 & -1 & -1 & 8 \\ 2 & 3 & -4 & \textcircled{0} \end{vmatrix}$$

Solution

The elements a_{11}, a_{23} and a_{44} are shown ringed in the given determinant. Their associated minors are found by omitting the rows and columns in which they occur. Hence,

$$\text{the minor of } a_{11} \text{ is} \quad \begin{vmatrix} 6 & -2 & 5 \\ -1 & -1 & 8 \\ 3 & -4 & 0 \end{vmatrix}$$

$$\text{the minor of } a_{23} \text{ is} \quad \begin{vmatrix} 1 & 7 & -3 \\ 0 & -1 & 8 \\ 2 & 3 & 0 \end{vmatrix}$$

$$\text{the minor of } a_{44} \text{ is} \quad \begin{vmatrix} 1 & 7 & 0 \\ 4 & 6 & -2 \\ 0 & -1 & -1 \end{vmatrix}$$

Notation

The minor of a_{ij} is denoted by $|M_{ij}|$. Notice that there are nine elements in a 3×3 matrix and therefore we can write down nine minors, one for each element. The *cofactor* of a_{ij} is the minor of a_{ij} multiplied by $(-1)^{i+j}$. This means that each minor is multiplied by $+1$ or -1 depending on whether $i + j$ is even or odd. In the 3×3 case the pattern of signs comes out as

$$\begin{bmatrix} + & - & + \\ - & + & - \\ + & - & + \end{bmatrix}$$

For example, in Example 2A, the cofactor of the element a_{23} is

$$-\begin{vmatrix} 1 & 7 & -3 \\ 0 & -1 & 8 \\ 2 & 3 & 0 \end{vmatrix}$$

and the cofactor of the element a_{44} is

$$(-1)^{4+4}|M_{44}| = +\begin{vmatrix} 1 & 7 & 0 \\ 4 & 6 & -2 \\ 0 & -1 & -1 \end{vmatrix}$$

Below we now give the formal definition of a cofactor of an element in a square matrix of any size.

Definition

The *cofactor* of a_{ij} is $(-1)^{i+j} |M_{ij}|$, where M_{ij} is the matrix obtained from the original one by deleting the ith row and the jth column. The cofactor of a_{ij} is written as $\mathrm{cof}(a_{ij})$.

Exercise 2D

Consider the 4×4 matrix

$$\begin{bmatrix} 1 & 2 & 3 & 4 \\ 0 & 1 & 0 & 1 \\ 1 & 0 & 1 & 0 \\ 0 & 2 & 3 & 0 \end{bmatrix}$$

From our definition all the cofactors must be 3×3 determinants. Write down the cofactors of a_{43}, a_{22}, a_{34} and evaluate them.

DERIVE ACTIVITY 2B

When you first load DERIVE it does not load all its inbuilt functions. There are many which are kept in UTILITY FILES. (See Chapter 9 in the DERIVE manual.) One of these utility files is VECTOR.MTH. We will need this more and more as we progress. To load it is quite straightforward.

(A)   From the **Author** line type T for **Transfer**, then L for **Load**, and finally U for **Utility**.
         At this stage you should have the line

         TRANSFER LOAD UTILITY file:

followed by the flashing cursor.
         Type in VECTOR.MTH and the relevant lines will be loaded while you wait. When it is fully loaded you will be returned to the author line again.
         Now we are ready to check our hand calculations. Input the 4×4 matrix in Exercise 2D on line 1.
         Now **Author**

         COFACTOR (#1, 1, 2)

Line 2 of Figure 2.2 should now be on your screen.
         Finally, **Simplify** and the full screen as shown in Figure 2.2 should be visible, verifying our answer.

$$\#2: \quad \text{COFACTOR}\left(\begin{bmatrix} 1 & 2 & 3 & 4 \\ 0 & 1 & 0 & 1 \\ 1 & 0 & 1 & 0 \\ 0 & 2 & 3 & 0 \end{bmatrix}, 1, 2\right)$$

$\#3: \quad \blacksquare$

```
COMMAND: author Build Calculus Declare Expand Factor Help Jump soLve Manage
         Options Plot Quit Remove Simplify Transfer Unremove moVe Window approX
Compute time: 0.1 seconds
Simp(#2)                              Free:89%              Derive Algebra
```

Figure 2.2 Finding the cofactor

The function **COFACTOR**(A,i,j) takes the matrix A, in our case #1, and finds $\text{cof}(a_{ij})$. Using this function verify the three other cofactors asked for in Exercise 2D.

(B) Before we leave this activity we introduce another function from the utility file VECTOR.MTH. **Author** and **Simplify**.

MINOR(#1, 1, 2)

This function gives you the minor of a_{12}. Compare the minor and cofactor of a_{12}.

This should verify the general result

$$\text{cof}(a_{ij}) = (-1)^{i+j} \, \text{minor}(a_{ij})$$

In our example the cofactor of a_{12} is -3 (line 3 in Figure 2.2), and the minor is 3. We can now write the determinant of a matrix in terms of its cofactors. For example, the general 3×3 case:

$$\text{DET}(A) = a_{11} \, \text{cof}(a_{11}) + a_{12} \, \text{cof}(a_{12}) + a_{13} \, \text{cof}(a_{13})$$

is Case 1 on page 25.

This is called evaluation (or expanding) by the first row. This form of determinant is very useful not only for evaluating determinants but for proving results in determinant theory. We can write down five other such expressions using different rows or columns. For example, expanding by the second column gives

$$\text{DET}(A) = a_{12} \, \text{cof}(a_{12}) + a_{22} \, \text{cof}(a_{22}) + a_{32} \, \text{cof}(a_{32})$$

Exercise 2E

For each of the following determinants write down the cofactors for the elements in the third row:

$$\begin{vmatrix} 1 & 0 & 0 \\ 2 & 3 & 6 \\ -1 & 4 & -1 \end{vmatrix} \qquad \begin{vmatrix} 0 & 1 & 2 \\ -1 & 2 & 0 \\ 3 & 4 & -2 \end{vmatrix} \qquad \begin{vmatrix} 5 & -1 & 3 \\ 2 & -2 & 1 \\ 4 & 0 & 7 \end{vmatrix}$$

Hence find the value of each determinant.

2.4   SOME RESULTS IN DETERMINANT THEORY

In this section we look at ways of evaluating determinants with the minimum of manipulation and calculation. We start with a seemingly obvious property of a determinant but it will turn out to be very useful in practice. Also, we can assume that the matrix is $n \times n$ for each of the properties listed.

(i)   If a determinant has a row (or column) in which all its entries are zero, then the value of that determinant must be zero. The proof of this statement is by expanding the determinant using the row or column containing the zeros.

(ii)   The determinant of a matrix is equal to the determinant of its transpose, that is, $\text{DET}(A) = \text{DET}(A^T)$.

(iii)   If we interchange two rows (or columns) in a determinant, then the sign of the determinant changes.

       An example in the 2×2 case (interchanging columns) would be

$$\text{DET}(A) = \begin{vmatrix} 1 & 2 \\ 3 & 4 \end{vmatrix} = 4 - 6 = -2$$

Interchanging the two columns gives

$$\begin{vmatrix} 2 & 1 \\ 4 & 3 \end{vmatrix} = 6 - 4 = 2 = -\text{DET}(A)$$

(iv)   Factorizing a determinant is different from factorizing a matrix. Recall that factorizing a matrix means that every element in the matrix has that factor. For a determinant, if a single row or column has a factor, then that factor can be taken out and placed before the determinant.

An example in the 2×2 case would be the following:

$$\begin{vmatrix} 10 & 25 \\ 30 & 27 \end{vmatrix} = 10\begin{vmatrix} 1 & 25 \\ 3 & 27 \end{vmatrix} = 10 \times 3\begin{vmatrix} 1 & 25 \\ 1 & 9 \end{vmatrix} = 30(9 - 25) = -480$$

Here we observe the factor 10 in the first column and then having factorized we observe a factor 3 in the second row. We end up with a determinant which cannot be factorized further and so we evaluate it. In larger square matrices this property can reduce large entries to manageable proportions.

(v) Probably the most useful property of a determinant is the following. To any row (or column) we can add a scalar multiple of any other row (or column) and the value of the determinant is unchanged.

As an example in the 2×2 case consider the following:

$$\begin{vmatrix} 25 & 26 \\ 4 & 3 \end{vmatrix} = \begin{vmatrix} 1 & 8 \\ 4 & 3 \end{vmatrix}$$

Here we have subtracted six lots of row 2 from row 1, or equivalently, have added –6 lots of row 2 to row 1. This now evaluates to $3 - 32 = -29$. Checking with the original matrix we see that $3 \times 25 - 4 \times 26 = 75 - 104 = -29$ again.

Example 2B

For a 3×3 matrix prove that if each entry in a single row or column has a factor b, then the determinant has a factor b.

Solution

Assume that there is a common factor b in the first row. Suppose that we expand the determinant in the cofactor form about the first row:

$$\text{DET}(A) = a_{11} \text{cof}(a_{11}) + a_{12} \text{cof}(a_{12}) + a_{13} \text{cof}(a_{13})$$

Since b is a common factor we can write

$$\text{DET}(A) = b\left[\frac{a_{11}}{b} \text{cof}(a_{11}) + \frac{a_{12}}{b} \text{cof}(a_{12}) + \frac{a_{13}}{b} \text{cof}(a_{13}) \right]$$

and each ratio $\dfrac{a_{11}}{b}$, $\dfrac{a_{12}}{b}$, $\dfrac{a_{13}}{b}$ is a whole number. Hence the result is proved.

Not all the facts given above are as easy to prove, but most of them begin with the expansion in cofactors. You are urged to try to prove some of the results given above.

DERIVE ACTIVITY 2C

In this activity we are going to explore how DERIVE can handle abstract matrices and determinants to at least check some of the properties mentioned in this chapter.

(A)   Input the matrix

$$\begin{bmatrix} a & b & c \\ d & e & f \\ g & h & i \end{bmatrix}$$

Now work out the determinant of this matrix using DET(#1) as before. Observe that DERIVE evaluates the determinant using the first row. Furthermore, the middle term is

$$+b(fg - di)$$

This is not the natural way of cofactors but, of course, the six terms we obtain when we expand are the same six terms we had earlier. Use **Expand** to see these six terms.

You should now have the screen as shown in Figure 2.3.

To obtain the expressions in this chapter simply replace $a, b, c, \ldots$ with $a_{11}, a_{12}, a_{13}, \ldots$ respectively.

#1: $\begin{bmatrix} a & b & c \\ d & e & f \\ g & h & i \end{bmatrix}$

#2:   DET $\begin{bmatrix} a & b & c \\ d & e & f \\ g & h & i \end{bmatrix}$

#3: $a \cdot (e \cdot i - f \cdot h) + b \cdot (f \cdot g - d \cdot i) + c \cdot (d \cdot h - e \cdot g)$

#4: $a\,e\,i - a\,f\,h - b\,d\,i + b\,f\,g + c\,d\,h - c\,e\,g$

COMMAND: **Author** Build Calculus Declare Expand Factor Help Jump soLve Manage
              Options Plot Quit Remove Simplify Transfer Unremove moVe Window approX
Compute time: 0.0 seconds
Expd(#3)   Free:89%   Derive Algebra

Figure 2.3   DERIVE working with general matrices

(B)   Now let us check the property given in 2.4 (v) using DERIVE.
       Author the matrix

$$\begin{bmatrix} a+dx & b+ex & c+fx \\ d & e & f \\ g & h & i \end{bmatrix}$$

Observe that we have taken the original matrix in this activity and replaced the first row with the first row plus x lots of the second row.

       We could have used columns or other rows to see this property at work. You are strongly urged to try more examples for yourself.

       Now evaluate the determinant of this matrix in the usual way. Your screen should look like that shown in Figure 2.4.

#1: $\begin{bmatrix} a + d\cdot x & b + e\cdot x & c + f\cdot x \\ d & e & f \\ g & h & i \end{bmatrix}$

#2: DET $\begin{bmatrix} a + d\cdot x & b + e\cdot x & c + f\cdot x \\ d & e & f \\ g & h & i \end{bmatrix}$

#3:   a (e i - f h) - b (f g - d i) + c (d h - e g)

COMMAND: **Author** Build Calculus Declare Expand Factor Help Jump soLve Manage
         Options Plot Quit Remove Simplify Transfer Unremove moVe Window approX
Compute time: 0.1 seconds
Simp(#2)   Free:89%   Derive Algebra

Figure 2.4   Demonstration of property 2.4(v)

From this you can see that the value of the determinant is the same as that obtained from the original matrix in Figure 2.3. This at least verifies the property stated in 2.4 (v).

(C)   Next let us verify the property in 2.4 (iii).
       Author the original matrix and also the matrix

$$\begin{bmatrix} d & e & f \\ a & b & c \\ g & h & i \end{bmatrix}$$

This matrix is obtained from the original by interchanging row 1 and row 2. Now work out the determinant of the original matrix and simplify.

Now work out the determinant of the matrix above and simplify.
To see the result author #3 + #5 and the result will be on line 7.
Ask DERIVE to simplify line 7 and it gives zero on line 8. This means

$$DET(A) + DET(B) = 0$$

This implies that DET(A) and DET(B) are negatives of each other and thus our proposition is verified.

The details of this activity are shown in Figure 2.5. Note that not every line will be visible on your screen at any one time.

$$
\#1: \quad \begin{bmatrix} a & b & c \\ d & e & f \\ g & h & i \end{bmatrix}
$$

$$
\#2: \quad \begin{bmatrix} d & e & f \\ a & b & c \\ g & h & i \end{bmatrix}
$$

$$
\#3: \quad DET \begin{bmatrix} a & b & c \\ d & e & f \\ g & h & i \end{bmatrix}
$$

$$
\#4: \quad a \cdot (e \cdot i - f \cdot h) + b \cdot (f \cdot g - d \cdot i) + c \cdot (d \cdot h - e \cdot g)
$$

$$
\#5: \quad DET \begin{bmatrix} d & e & f \\ a & b & c \\ g & h & i \end{bmatrix}
$$

$$
\#6: \quad a \cdot (f \cdot h - e \cdot i) + b \cdot (d \cdot i - f \cdot g) + c \cdot (e \cdot g - d \cdot h)
$$

$$
\#7: \quad DET \begin{bmatrix} a & b & c \\ d & e & f \\ g & h & i \end{bmatrix} + DET \begin{bmatrix} d & e & f \\ a & b & c \\ g & h & i \end{bmatrix}
$$

#8: 0

```
COMMAND: Author Build Calculus Declare Expand Factor Help Jump soLve Manage
         Options Plot Quit Remove Simplify Transfer Unremove moVe Window approX
Enter option
Simp(#7)                                  Free:74%                  Derive Algebra
```

Figure 2.5   Interchanging rows changes the sign of the determinant

(D)   Finally in this activity we verify 2.4 (ii), that is, the determinant of a matrix is equal to the determinant of its transpose. Before you start you will need to recall how to put the transpose of a matrix into DERIVE.

First **Author** the matrix below in line 1:

$$\begin{bmatrix} a & b & c \\ d & e & f \\ g & h & i \end{bmatrix}$$

Now **Author** the transpose of this matrix in line 2. **Simplify** line 2 to obtain line 3, the transposed matrix.

Author the determinant of our original matrix in line 1 and simplify it. This gives us lines 4 and 5.

Finally, **Author** the determinant of line 3, the transposed matrix, and **Simplify** it.

Comparison of line 5 and line 7 shows that the determinant of the transpose is the same as the determinant of the original matrix.

Your screen should look like that shown in Figure 2.6.

#1: $\begin{bmatrix} a & b & c \\ d & e & f \\ g & h & i \end{bmatrix}$

#2: $\begin{bmatrix} a & b & c \\ d & e & f \\ g & h & i \end{bmatrix}$`

#3: $\begin{bmatrix} a & d & g \\ b & e & h \\ c & f & i \end{bmatrix}$

#4:   DET $\begin{bmatrix} a & b & c \\ d & e & f \\ g & h & i \end{bmatrix}$

#5: $a \cdot (e \cdot i - f \cdot h) + b \cdot (f \cdot g - d \cdot i) + c \cdot (d \cdot h - e \cdot g)$

#6:   DET $\begin{bmatrix} a & d & g \\ b & e & h \\ c & f & i \end{bmatrix}$

#7: $a \cdot (e \cdot i - f \cdot h) + b \cdot (f \cdot g - d \cdot i) + c \cdot (d \cdot h - e \cdot g)$

COMMAND: **Author** Build Calculus Declare Expand Factor Help Jump soLve Manage
         Options Plot Quit Remove Simplify Transfer Unremove moVe Window approX
Compute time: 0.0 seconds
Simp(#6)   Free:74%   Derive Algebra

Figure 2.6   The determinant of the transpose equals the determinant of the
              matrix

Further Properties of Determinants

Here let A and B be two square matrices of the same size.

1. $\text{DET}(A \cdot B) = \text{DET}(A)\,\text{DET}(B)$
       This property will be proved in Chapter 4, but will be used before then.

2. $\text{DET}(A^{-1}) = (\text{DET}(A))^{-1} = \dfrac{1}{\text{DET}(A)}$

       We can prove this fact from (1) above by choosing the matrix B as A^{-1}. We now have

$$\text{DET}(A \cdot A^{-1}) = \text{DET}(A) \cdot \text{DET}(A^{-1})$$

The left-hand side of this equation simplifies to $\text{DET}(I)$ and its value is 1. Hence,

$$1 = \text{DET}(A) \cdot \text{DET}(A^{-1})$$

Because A^{-1} exists, $\text{DET}(A) \neq 0$. The result follows by dividing both sides of the above equation by $\text{DET}(A)$.

Consolidation Exercise

1.   Evaluate the following determinants by hand, using properties of determinants. Check your answer using DERIVE:

    (i) $\begin{vmatrix} 1 & -1 & 2 \\ 2 & 1 & 1 \\ 1 & -3 & 1 \end{vmatrix}$ (ii) $\begin{vmatrix} 2 & 3 & -5 \\ 1 & 7 & -2 \\ 5 & -11 & 2 \end{vmatrix}$

2.   Use Cramer's rule to solve the following systems of linear equations:

    (i)   (ii)

$$\begin{aligned} x - y + 2z &= 1 \\ 2x + y + z &= 2 \\ x - 3y + z &= 1 \end{aligned} \qquad\qquad \begin{aligned} 2x + 3y - 5z &= 4 \\ x + 7y - 2z &= 1 \\ 5x - 11y + 2z &= -2 \end{aligned}$$

3.   *Without* expanding the determinant fully, show that

$$\begin{bmatrix} a_1^2 & a_1 & 1 \\ a_2^2 & a_2 & 1 \\ a_3^2 & a_3 & 1 \end{bmatrix} = -(a_1 - a_2)(a_2 - a_3)(a_3 - a_1)$$

4.   Evaluate the following determinants:

(i) $\begin{vmatrix} 1 & 6 \\ 2 & 1 \end{vmatrix}$ (ii) $\begin{vmatrix} 3 & 1 & -1 \\ 2 & 0 & 4 \\ 1 & -5 & 1 \end{vmatrix}$ (iii) $\begin{vmatrix} 6 & -17 & 102 & 46 \\ 0 & 1 & 16 & -165 \\ 0 & 0 & 3 & -14 \\ 0 & 0 & 0 & 2 \end{vmatrix}$

5.   By hand, evaluate the following determinants more than once using different approaches and properties to find the shortest and neatest solution. Check your numerical value using DERIVE:

(i) $\begin{vmatrix} 17 & 46 & 7 \\ 20 & 49 & 8 \\ 23 & 52 & 9 \end{vmatrix}$ (ii) $\begin{vmatrix} 17 & 46 & 7 \\ 20 & 49 & 8 \\ 23 & 52 & 12 \end{vmatrix}$ (iii) $\begin{vmatrix} 1001 & 17 & 2 \\ 1002 & 18 & 8 \\ 1003 & 19 & 7 \end{vmatrix}$

6.   *Without* expanding the determinant fully, show that

$$\begin{vmatrix} 1 & a & a^2 & a^3 \\ a^3 & 1 & a & a^2 \\ b & a^3 & 1 & a \\ c & d & a^3 & 1 \end{vmatrix} = (1 - a^4)^3$$

7.   If A is an $n \times n$ matrix and α is any scalar, show that

$$\mathrm{DET}(\alpha A) = \alpha^n \mathrm{DET}(A)$$

3

Vectors

3.1 ROW AND COLUMN VECTORS

In this chapter we show how vectors are really examples of matrices. Hence, using the theory of Chapter 1 we can proceed to discuss vectors and their properties.

Definitions

A *row vector* is a $1 \times n$ matrix. It is sometimes referred to as an *n*-vector. A *column vector* is an $n \times 1$ matrix. It also is an *n*-vector but now we write it as a column.

For example, $A = [1\ 1\ 3]$ is a 3-vector (a row vector). It can be thought of as an ordered triple $(1, 1, 3)$. $A^T = \begin{bmatrix} 1 \\ 1 \\ 3 \end{bmatrix}$ is an example of a column vector.

Notation

In general an ordered triple is (x, y, z) where x, y and z belong to the real numbers, denoted $\mathfrak{R}$. We write $(x, y, z) \in \mathfrak{R}^3$, and understand this to mean 'the ordered triple (x, y, z) belongs to all possible ordered triples of three real numbers'.

The symbol '$\in$' means 'belongs to'.

$\mathfrak{R}^3$ is referred to as 3-space or *Euclidean three space*. In our example we could write $(1, 1, 3) \in \mathfrak{R}^3$. The triple $(1, 1, 3)$ can be thought of as the line joining $(0, 0, 0)$ (the origin) to $(1, 1, 3)$ in 3-space.

Compare the above ideas with those of 2-space or $\mathfrak{R}^2$, the set of ordered pairs. Every time we plot a graph with the two axes x and y, we are using 2-space and every point in this space is represented by an ordered pair of the form (x, y), where x and y are real numbers.

The definition for $\Re^2$ is the following:

$$\Re^2 = \{(x,y) : x \in \Re \text{ and } y \in \Re\}$$

In a similar way,

$$\Re^3 = \{(x, y, z) : x \in \Re, y \in \Re \text{ and } z \in \Re\}$$

In general, n-space or Euclidean n-space is defined as follows:

$$\Re^n = \{(x_1, x_2, x_3, ..., x_n) : x_i \in \Re \text{ for all } i\}$$

The objects in this set are often called 'ordered n-tuples', for the obvious reason!

DERIVE ACTIVITY 3A

In this activity you will learn how to input vectors.

(A)   First try **Author** (1,1,3). You should find that DERIVE will not accept it and points out that there is a syntax error.
　　　　　Now try {1,1,3}. The same thing should happen, rejection!
　　　　　Now try [1,1,3] and DERIVE should accept it.
　　　　　This may be a little surprising since although a row vector is a matrix and matrices are always enclosed in square brackets we do not put commas between the elements.

(B)   Alternatively, let us start from the beginning again with the **Declare** option but instead of **Matrix**, press R for **vectoR**. DERIVE will now ask you what the dimension of your vector is.   At this stage it is simply the number of coordinates your vector has. In this activity we have three coordinates. Now input the two 1s and a 3.
　　　　　The display should show [1,1,3]. Notice that the commas are inserted for you.
　　　　　If we had declared a matrix of size 1×3 we would have the display [1 1 3].
　　　　　As in Chapter 1 a matrix has no commas present. Spaces distinguish where one element finishes and the next one starts.
　　　　　In summary, DERIVE inputs a vector in coordinate form and therefore puts in the commas.

(C) An alternative definition for a matrix is now possible. A matrix is a vector of row vectors, each row vector having the same number of coordinates or elements.

$$\text{Input the matrix} \begin{bmatrix} 1 & 2 & 3 \\ 4 & 5 & 6 \\ 7 & 8 & 9 \end{bmatrix} \text{as you would have done in Chapter 1.}$$

Now we input this matrix as a vector of row vectors. We need to declare that we are dealing with row vectors and that there are three row vectors to input.
Press D for **Declare**, then R for **vectoR**, then 3 for the dimension, as we have three row vectors to input.

Now type [1,2,3] and press Enter. Now type [4,5,6] and press Enter. Finally, type [7,8,9] and press Enter and the vector of vectors should look like our original matrix above.

(D) Now input the 2 × 5 matrix $\begin{bmatrix} 1 & 2 & 3 & 4 & 5 \\ 5 & 4 & 3 & 2 & 1 \end{bmatrix}$ using **Declare vectoR**. Think about the dimension you must input before you start.

Further Notation

Another notation used to represent the row vector [1, 1, 3] is

$1\mathbf{i} + 1\mathbf{j} + 3\mathbf{k}$

where **i**, **j** and **k** are the ordered triples (1, 0, 0), (0, 1, 0) and (0, 0, 1) respectively:

$1(1, 0, 0) + 1(0, 1, 0) + 3(0, 0, 1) = (1, 1, 3)$

This addition is the same as the addition of matrices introduced in Chapter 1.

i can be thought of as the line joining (0, 0, 0) to (1, 0, 0) and will therefore be of length one (or unit) and it will be along the x axis in 3-space.

j, in a similar way, can be thought of as the unit vector along the y axis and **k** will be the unit vector along the z axis in 3-space.

DERIVE ACTIVITY 3B

For this activity you will need to load VECTOR.MTH as described in Chapter 2.

(A)   At the **Author** command input $i\_$ and press Enter.

Similarly, input $j\_$ and $k\_$ and press Enter after each one. Your screen should look like that shown in Figure 3.1.

```
#1:  i_

#2:  [1, 0, 0]

#3:  j_

#4:  [0, 1, 0]

#5:  k_

#6:  [0, 0, 1]
```

```
COMMAND: Author Build Calculus Declare Expand Factor Help Jump soLve Manage
         Options Plot Quit Remove Simplify Transfer Unremove moVe Window approX
Enter option
User                               Free:74%              Derive Algebra
```

Figure 3.1   The unit vectors **i**, **j** and **k** in DERIVE

(B)   **Author** $i\_ + j\_ + 3k\_$. What vector do you obtain?

Input the following vectors using $i\_, j\_$ and $k\_$:

$$(4, -1, 3) \qquad (6, -11, -17) \qquad (-1, 0, 0) \qquad (-2, 3, -5)$$

3.2   EQUALITY OF VECTORS

Because we can think of a vector as a line joining two points it clearly has a length (or norm or magnitude) and also a direction (usually measured relative to axes). Its length is obtained by an extension of Pythagoras' theorem to 3-space:

For the vector $\mathbf{a} = [x, y, z] \in \Re^3$ its length $|\mathbf{a}| = \sqrt{(x^2 + y^2 + z^2)}$.

Notice that the letter **a** is in bold to denote that we have a vector quantity. When writing vectors by hand we usually underline the symbol, that is, we write a̲.

Remember, |a| denotes the length of the vector, that is, the length in Euclidean 3-space from $(0, 0, 0)$ to (x, y, z). Also in some texts, ‖a‖ is used for norm (or length).

Example 3A

Find the length of the vector $\mathbf{a} = [1, 1, 3]$.

Solution

The length of the vector $\mathbf{a} = [1, 1, 3]$, is $\sqrt{(1^2 + 1^2 + 3^3)} = \sqrt{11}$, using the Pythagorean formula given above.

$$\text{We write } |\mathbf{a}| = \sqrt{11}.$$

Example 3B

The general formula for the distance between the two points (x_1, y_1, z_1) and (x_2, y_2, z_2) is given by

$$\sqrt{((x_1 - x_2)^2 + (y_1 - y_2)^2 + (z_1 - z_2)^2)}$$

Use this formula to find the length of the vector starting at $(7, 8, 1)$ and finishing at $(8, 9, 4)$.

Solution

Using the formula, the length of the vector is

$$\sqrt{(7-8)^2 + (8-9)^2 + (1-4)^2} = \sqrt{11}$$

If we denote the vector in Example 3B by $\mathbf{b}$, then $|\mathbf{a}| = |\mathbf{b}| = \sqrt{11}$. This means that the lengths of $\mathbf{a}$ and $\mathbf{b}$ are equal.

     Further, their directions are also the same, that is, they are parallel to each other, since their starting points and finishing points are related by the movements one unit parallel to the x axis, one unit parallel to the y axis and three units parallel to the z axis. Hence we say that vector $\mathbf{a}$ equals the vector $\mathbf{b}$.

$$\text{We write } \mathbf{a} = \mathbf{b}.$$

Definition

> Two vectors are *equal* if they have the same length and the same direction.

DERIVE ACTIVITY 3C

Again you will need to load VECTOR.MTH before starting this activity.

(A)   **Author** $1i\_ + 1j\_ + 1k\_$.

Now to find the length or norm of this vector we ask for the absolute value of line 2 by **Author** ABS(#2).

Finally, **Simplify** this expression to obtain line 4 in Figure 3.2.

Now input a general vector $ai\_ + bj\_ + ck\_$ on line 5 and obtain its length in the same way as we obtained the length of $1i\_ + 1j\_ + 1k\_$ above. Your screen should look like that shown in Figure 3.2.

#1: $1 \cdot i\_ + 1 \cdot j\_ + 1 \cdot k\_$

#2:   [1, 1, 1]

#3: |[1, 1, 1]|

#4:   √3

#5: $a \cdot i\_ + b \cdot j\_ + c \cdot k\_$

#6: $|a \cdot i\_ + b \cdot j\_ + c \cdot k\_|$

#7: $\sqrt{(a^2 + b^2 + c^2)}$

COMMAND: **Author** Build Calculus Declare Expand Factor Help Jump soLve Manage
         Options Plot Quit Remove Simplify Transfer Unremove moVe Window approX
Compute time: 0.1 seconds
Simp(#6)   Free:74%   Derive Algebra

Figure 3.2   Modulus of vectors using ABS

Exercise 3A

Find the length of each of the following vectors.   Which vectors have the same length?

[1, 0, 1]   [2, –1, 3]   [1, 2, –3]   [–2, 1, –3]

[3, –2, 0]   [1, 1, 1]   [3, 0, –3]   [3, 4, 7]

3.3   ADDITION AND MULTIPLICATION OF VECTORS

Addition

Because we can think of vectors as matrices, we can add vectors as we added matrices in Chapter 1.

Exercise 3B

In matrix notation we have, if $\mathbf{a} = [1\ 1\ 3]$ and $\mathbf{b} = [0\ {-}3\ 9]$ then $\mathbf{a} + \mathbf{b} = [1\ {-}2\ 12]$.

Write this line down again in coordinate or vector form.

Work out $\mathbf{a} - \mathbf{b}$ giving your answer in both matrix form and coordinate form.

Work out $\mathbf{b} + \mathbf{a}$.

We say vectors *commute under addition* if $\mathbf{a} + \mathbf{b} = \mathbf{b} + \mathbf{a}$ for any conformable pair of vectors $\mathbf{a}$ and $\mathbf{b}$.

From your work in this exercise, show that the vectors $\mathbf{a}$ and $\mathbf{b}$ commute under addition?

Prove in general that vectors commute under addition. Do vectors commute under subtraction?

Multiplication

For multiplication we have more than one product of two vectors, $\mathbf{u}$ and $\mathbf{v}$. We begin with the simplest one, known in the literature as the 'dot product' or the 'scalar product' (the dot product because we write $\mathbf{u}{\cdot}\mathbf{v}$, the scalar product because the result of the multiplication is always a scalar).

Exercise 3C

Let $\mathbf{u} = [x_1\ y_1\ z_1]$ and $\mathbf{v} = [x_2\ y_2\ z_2]$ be two vectors in matrix form. Write down $\mathbf{v}^T$ and then work out the matrix product $\mathbf{u}{\cdot}\mathbf{v}^T$. Your working should look similar to that given below:

$$[x_1\ y_1\ z_1]\cdot\begin{bmatrix} x_2 \\ y_2 \\ z_2 \end{bmatrix} = x_1 x_2 + y_1 y_2 + z_1 z_2$$

Notice that the sizes make the matrices conformable under multiplication:

$$(1 \times 3) \cdot (3 \times 1) = (1 \times 1)$$

Observe that the right-hand side of this equation is indeed a scalar, a 1×1 matrix.

If the x_i, y_i and z_i values are all chosen from the real numbers $\Re$, then the product is also a number in $\Re$.

Scalar Product

Definition

> The *scalar product* of two row vectors, **u** and **v**, is written as **u·v** and it is computed as **u·v**^{T} for conformability.

Observe the fact that **v·u** = **u·v**, because $x_1 x_2 = x_2 x_1$, etc., in $\Re$ and thus the scalar product is commutative. Keep Exercise 3C in mind as you read the above definition. Recall how in Chapter 1 we multiplied two matrices together:

$$\begin{bmatrix} 1 & 2 \\ 3 & 4 \end{bmatrix} \cdot \begin{bmatrix} 2 & 3 & 1 \\ 1 & 0 & 2 \end{bmatrix} = \begin{bmatrix} (1 \times 2 + 2 \times 1) & (1 \times 3 + 2 \times 0) & (1 \times 1 + 2 \times 2) \\ (3 \times 2 + 4 \times 1) & (3 \times 3 + 4 \times 0) & (3 \times 1 + 4 \times 2) \end{bmatrix}$$

All the products are scalar products in the sense that we have defined above.

We have worked out $[1 \quad 2] \cdot \begin{bmatrix} 2 \\ 1 \end{bmatrix}$ and $[1 \quad 2] \cdot \begin{bmatrix} 3 \\ 0 \end{bmatrix}$ and similar expressions to obtain the entries in the product matrix. We have now given a reason for the necessity of a dot in matrix multiplication in Chapter 1 and how DERIVE sees matrix multiplication.

3.4   A GEOMETRICAL ALTERNATIVE

There is an alternative definition for the scalar product of a more geometrical flavour and because the two definitions must be equivalent we obtain some useful equations.

Consider again the two vectors, **u** = $[x_1 \ y_1 \ z_1]$ and **v** = $[x_2 \ y_2 \ z_2]$, visualized starting at the origin (0, 0, 0) and ending at (x_1, y_1, z_1) and (x_2, y_2, z_2), respectively, in Euclidean 3-space. Let the angle between them be θ.

Definition

> The scalar product of two vectors **u** and **v** is defined by
>
> **u·v** = $|$**u**$|$ $|$**v**$|$ $\cos \theta$

Notice again this is a scalar quantity, being the product of two lengths and the cosine of an angle. Equating our two definitions, we have the following equation:

$$|\mathbf{u}||\mathbf{v}|\cos(\theta) = x_1 x_2 + y_1 y_2 + z_1 z_2 \qquad (3.1)$$

Exercise 3D

1. Draw a triangle with sides of three lengths, $|\mathbf{u}|$, $|\mathbf{v}|$ and $|\mathbf{u} - \mathbf{v}|$ and label the angle θ between $|\mathbf{u}|$ and $|\mathbf{v}|$.

    Now recall the cosine formula for a triangle of sides a, b and c, with θ being the angle between sides b and c:

    $$a^2 = b^2 + c^2 - 2bc \cdot \cos(\theta)$$

    Show that the result given above in Equation (3.1) comes directly from the cosine formula.

2. Calculate all possible scalar products using the following vectors:

    $$\mathbf{a} = (1, 0, 1), \quad \mathbf{b} = (2, -1, 3), \quad \mathbf{c} = (-1, 1, 4), \quad \mathbf{d} = (1, 1, 1)$$

Hints to get you started and to get you through this exercise

1. Label the vertex where the angle θ is marked as the origin $(0,0)$ and the other vertices as $\mathbf{u} = (x_1, y_1)$ and $\mathbf{v} = (x_2, y_2)$.
2. Put $|\mathbf{u}| = b$, $|\mathbf{v}| = c$ and $|\mathbf{u} - \mathbf{v}| = a$, so that θ is between b and c.
3. Observe that the left-hand side of (3.1) appears in one of the terms in your equation if your substitution is correct.
4. Expand $|\mathbf{u} - \mathbf{v}|^2$, $|\mathbf{u}|$ and $|\mathbf{v}|$, in terms of x_1, x_2, y_1, and y_2, using a reduced version of the formula given earlier in this chapter for the distance between two points.
5. Cancel terms from both sides of your equation to obtain the result.

3.5 ORTHOGONALITY OF VECTORS

Definition

Two vectors are said to be *orthogonal* if they are at right angles to each other. From this definition and the above discussion of scalar products we see that we have a criterion for when two vectors are orthogonal.

> Two vectors are *orthogonal* if their scalar product is zero.

This definition is consistent, for if $\theta = 90°$ then $\cos(\theta) = 0$. Notice that the converse is not necessarily true. If the scalar product of two vectors is zero are the two vectors necessarily at right angles, or orthogonal?

There are three options: either $\mathbf{a} = \mathbf{0}$ or $\mathbf{b} = \mathbf{0}$ or $\cos\theta = 0$.

Exercise 3E

Consider the two vectors in $\mathfrak{R}^3$, $\mathbf{u} = [1, 2, -1]$ and $\mathbf{v} = [2, 1, 4]$. Show that these two vectors are orthogonal.

Find a vector in $\mathfrak{R}^3$ which is orthogonal to both the above vectors.

3.6   THE VECTOR PRODUCT IN $\mathfrak{R}^3$

As the name suggests, this product when defined should result in a vector rather than a scalar. The vector product will be denoted by $\mathbf{u} \times \mathbf{v}$ and is also called the cross product because of the sign between the vectors.

Definition

Let $\mathbf{u}$ and $\mathbf{v}$ be the vectors $[x_1, y_1, z_1]$ and $[x_2, y_2, z_2]$, respectively, then we define the *vector product* by a 3×3 determinant as follows:

$$\mathbf{u} \times \mathbf{v} = \begin{vmatrix} \mathbf{i} & \mathbf{j} & \mathbf{k} \\ x_1 & y_1 & z_1 \\ x_2 & y_2 & z_2 \end{vmatrix} \tag{3.2}$$

As before, there is a second way of defining this product analogous to the alternative definition of the scalar product.

Alternative Definition

$$\mathbf{u} \times \mathbf{v} = |\mathbf{u}||\mathbf{v}|\sin(\theta)\mathbf{n}$$

where $\mathbf{n}$ is a unit vector orthogonal to the plane containing $\mathbf{u}$ and $\mathbf{v}$ and having the same direction as the translation of a right-handed screw due to a rotation from $\mathbf{u}$ to $\mathbf{v}$. It follows that the magnitude of the vector given in (3.2) above is just

$$|\mathbf{u}||\mathbf{v}|\sin(\theta)$$

It follows from the properties of determinants that

$$\mathbf{u} \times \mathbf{v} \neq \mathbf{v} \times \mathbf{u}$$

but

$$\mathbf{u} \times \mathbf{v} = -\mathbf{v} \times \mathbf{u}$$

Another property of vector product is that it is geometrical. The magnitude of the vector product **u** × **v** is the same as the area of the parallelogram formed from the two vectors **u** and **v**. Even though vector products themselves are often said to be difficult to handle, this property can be seen when we recall the area of a parallelogram given two sides and the included angle. This product has many applications in physics and engineering.

Exercise 3F

For the following vectors:

a = [1, –1, 1]   **b** = [0, 1, 3]   **c** = [–1, 2, 4]

d = [2, 0, 3]   **e** = [–1, 1, –1]

calculate the vector products

a × **b**,   **b** × **c**,   **d** × **e**,   **a** × **e**,   **b** × **a**,   **c** × **b**

Verify the property **u** × **v** = –**v** × **u**.
What can you deduce about vectors **a** and **e**?

DERIVE ACTIVITY 3D

Scalar or dot products are the 'usual products' that DERIVE uses in matrix multiplications. The cross or vector product is an inbuilt function of DERIVE. It should be clear from our definition that the vectors in the cross product must have the same number of entries. We cannot cross the vector [1, 2] with [1, 2, 3] but we can cross the vector [1, 2, 3] with the vector [–1, 5, 0].

(A)   Load DERIVE and then **Author** the two vectors [1, 2, 3] and [–1, 5, 0] on lines 1 and 2 respectively.
        Now **Author**

        CROSS(#1,#2)

and then **Simplify**. You should have the screen shown in Figure 3.3.

```
#1:   [1, 2, 3]

#2:   [-1, 5, 0]

#3:   CROSS([1, 2, 3], [-1, 5, 0])

#4:   [-15, -3, 7]
```

```
COMMAND: Author Build Calculus Declare Expand Factor Help Jump soLve Manage
         Options Plot Quit Remove Simplify Transfer Unremove moVe Window approX
Compute time: 0.1 seconds
Simp(#3)                                Free:46%                    Derive Algebra
```

Figure 3.3   The vector product of two vectors using CROSS

If we had done this exercise by hand we would have had the following working.
Let **u** = [1, 2, 3] and **v** = [-1, 5, 0]. By definition,

$$\mathbf{u} \times \mathbf{v} = \begin{vmatrix} \mathbf{i} & \mathbf{j} & \mathbf{k} \\ 1 & 2 & 3 \\ -1 & 5 & 0 \end{vmatrix}$$

Now expanding this determinant by the first row we have

$$(2 \times 0 - 3 \times 5)\mathbf{i} - (1 \times 0 - (-1) \times 3)\mathbf{j} + (1 \times 5 - (-1) \times 2)\mathbf{k}$$

When simplified this gives us

$$-15\mathbf{i} + (-3)\mathbf{j} + 7\mathbf{k}$$

This is simply the fourth line of Figure 3.3 in the **i**, **j**, **k** notation.

(B)   Use DERIVE to calculate the cross products:

(i) $\mathbf{i} \times \mathbf{i}$,   (ii) $\mathbf{j} \times \mathbf{j}$,   (iii) $\mathbf{k} \times \mathbf{k}$,   (iv) $\mathbf{v} \times \mathbf{u}$

(using the vectors given in (A)). Verify from your results, the property $\mathbf{u} \times \mathbf{v} = -\mathbf{v} \times \mathbf{u}$.

(C)   Use DERIVE to calculate the dot product and cross product of each pair of the following vectors:

$$(1, 0, 1) \qquad (-1, 2, 3) \qquad (4, -1, 2) \qquad (2, -4, -6)$$

Which pair of vectors is orthogonal? Which pair of vectors is parallel?

3.7   APPLICATIONS TO GEOMETRY

Vectors can make many geometrical theorems easier to prove. Take, for example, the fact that the line joining the mid-points of two sides of a triangle is parallel to the third side and equal to half its length.

Exercise 3G

Draw a triangle with vertices OAB and join the midpoints of OA and OB. Label the mid-point of OA, X and that of OB, Y. Let $\overrightarrow{OA} = \mathbf{a}$ and $\overrightarrow{OB} = \mathbf{b}$. Then

$$\overrightarrow{OX} = \mathbf{a}/2 \quad \text{and} \quad \overrightarrow{OY} = \mathbf{b}/2 \quad \text{and} \quad \overrightarrow{AB} = \mathbf{b} - \mathbf{a} \tag{3.3}$$

Now consider

$$\overrightarrow{XY} = \overrightarrow{XO} + \overrightarrow{OY} = -\mathbf{a}/2 + \mathbf{b}/2 = (\mathbf{b} - \mathbf{a})/2 \tag{3.4}$$

Hence the result follows by comparing (3.3) and (3.4). The two vectors are parallel and

$$\left| \overrightarrow{XY} \right| = \frac{\left| \overrightarrow{AB} \right|}{2}$$

Exercise 3H

In this exercise we will prove the fact that the diagonals of a parallelogram bisect each other.
        Draw a parallelogram with vertices A, B, C and D, with AB parallel to CD. Let X be the midpoint of AC and Y be the midpoint of BD. Now let A, B, C, D, X, Y have position vectors $\mathbf{a}$, $\mathbf{b}$, $\mathbf{c}$, $\mathbf{d}$, $\mathbf{x}$, $\mathbf{y}$ respectively. Then $\mathbf{x} = \frac{1}{2}(\mathbf{a} + \mathbf{c})$ and $\mathbf{y} = \frac{1}{2}(\mathbf{b} + \mathbf{d})$.
        Now $BC = AD$ and BC is parallel to AD and thus $\mathbf{c} - \mathbf{b} = \mathbf{d} - \mathbf{a}$. This equation is equivalent to

$$\mathbf{c} + \mathbf{a} = \mathbf{b} + \mathbf{d}$$

Thus $\mathbf{x} = \mathbf{y}$ and hence X and Y are the same point.
        The next example makes use of the scalar product to find the acute angle between two vectors.

Example 3C

Find the cosine of the angle between the vectors [1, 1, 1] and [1, 2, 3].

Solution

Here we recall that $a \cdot b = |a||b| \cos(\theta)$ and for the vectors $a = [1, 1, 1]$ and $b = [1, 2, 3]$:

$$a \cdot b = 1 \times 1 + 1 \times 2 + 1 \times 3 = 6$$

Furthermore, $|a| = \sqrt{(1^2 + 1^2 + 1^2)} = \sqrt{3}$ and $|b| = \sqrt{(1^2 + 2^2 + 3^2)} = \sqrt{14}$. Hence we have $\cos(\theta) = 6/(\sqrt{3} \times \sqrt{14}) = 6/\sqrt{42}$

The acute angle between vector a and b is 0.388 radians or 22.2°.

Exercise 3I

Find the acute angle between the vector $a = [1, 1, 1]$ and the following vectors:

$$p = [-1, 2, 1] \qquad r = [3, -2, 2] \qquad s = [-2, -1, 0] \qquad t = [2, 3, -6]$$

3.8   LINEAR DEPENDENCE AND INDEPENDENCE OF VECTORS

In what follows we will use v_1, v_2, v_3, ... to represent vectors and α_1, α_2, α_3, ... to represent scalars.

Definitions

If two vectors are related by an equation of the form

$$v_1 = \alpha v_2$$

then we say that v_1 *is a scalar multiple of* v_2. Further, we say that v_1 *depends on* v_2 and vice versa.

The set of vectors $\{v_1, v_2\}$ is said to be a *dependent* set. This idea can extend to any number of vectors but it will need formalizing for it to be useful. Let

$$V = \{v_1, v_2, v_3, v_4, ..., v_n\}$$

be a set of n vectors.

Consider the sum of vectors multiplied by scalars given below:

$$\alpha_1 v_1 + \alpha_2 v_2 + a_3 v_3 + \alpha_4 v_4 + \ldots + \alpha_n v_n$$

We call this sum a *linear combination* of the vectors v_i.

Example 3D

Consider the three vectors in $\Re^4$ given below:

$$v_1 = [1, 2, 3, 4], \quad v_2 = [1, 3, 6, 9], \quad \text{and} \quad v_3 = [2, 4, 6, 8]$$

Show that v_1 and v_3 are dependent and that v_1 and v_2 are independent.

Solution

By inspection,

$$v_3 = 2v_1$$

Hence v_1 and v_3 are dependent by the definition given above.

In the same way v_1 and v_2 are independent since $v_1 \neq \alpha v_2$ for any scalar α. The set $\{v_1, v_2, v_3\}$ is thus a dependent set, whereas the subset $\{v_1, v_2\}$ is an independent set.

Exercise 3J

1.   Consider the three vectors in $\Re^6$ given by

$$v_1 = [-1, 2, 1, 0, 3, 4], \ v_2 = [6, 1, -3, 2, 0, -1], \ v_3 = [2, -4, -2, 0, -6, -8]$$

      Show that v_1 and v_3 form a dependent set and that v_2 and v_3 are independent. Deduce that the set $\{v_1, v_2, v_3\}$ forms a dependent set of vectors.

2.   If $V = \{v_1, v_2, v_3, v_4\}$ is a set of four vectors and $\alpha_1, \alpha_2, \alpha_3, \alpha_4$ are scalars, which of the following is a linear combination of vectors?

      (i) $\alpha_1^2 v_1 + \alpha_2 v_2$

      (ii) $\alpha_1 \alpha_3 v_1 + \alpha_2 v_2 + \alpha_3 v_3 + v_4$

      (iii) $\alpha_1 v_1 \cdot v_2 + \alpha_3 v_3 + \alpha_4 v_4$

      (iv) $\alpha_1 v_1 + \alpha_4 v_1 \times v_4$

We now motivate the formal definitions of dependence and independence by considering linear combinations and generalizing the above ideas.

If $v_1 = \alpha v_2$ then $v_1 - \alpha v_2 = 0$ (the zero vector). Hence, if we have the more general equation

$$\alpha_1 v_1 + \alpha_2 v_2 + \alpha_3 v_3 + ... + \alpha_n v_n = 0 \qquad (3.5)$$

with one or more of the αs not equal to zero, we must have a dependent set of vectors $\{v_1, v_2, v_3, ..., v_n\}$ as we will now show.

Let us assume that one of the αs is not zero. Without loss of generality we choose $\alpha_j \neq 0$.

Now we rearrange Equation (3.5) to make v_j the subject of the formula. We obtain the equivalent equation

$$v_j = -\frac{1}{\alpha_j}(\alpha_1 v_1 + \alpha_2 v_2 + \alpha_3 v_3 + ... + \alpha_{j-1} v_{j-1} + \alpha_{j+1} v_{j+1} + ... + \alpha_n v_n)$$

showing that v_j *depends* on some or all of the remaining vectors in our set. From this work we can obtain, by negation, a definition for independence of a set of vectors.

Definition

A set of vectors $\{v_1, v_2, v_3, ..., v_n\}$ is said to be *linearly independent* (or a *frame*) if and only if the equation

$$\alpha_1 v_1 + \alpha_2 v_2 + \alpha_3 v_3 + ... + \alpha_n v_n = 0$$

implies $\alpha_i = 0$ for all i values.

Note that in this definition we have two zeros, the first one representing the zero vector and the second one representing the zero scalar.

Example 3E

Consider the set of vectors $\{v_1, v_2, v_3\}$ in $\Re^3$ given by

$$v_1 = (1, 1, 1), \quad v_2 = (1, -2, 1) \quad \text{and} \quad v_3 = (7, -11, 7)$$

Is this set a frame (i.e. a set of independent vectors) or is it a dependent set of vectors?

Solution

We start by taking a linear combination of the three vectors given and then equating this combination to the zero vector as we did in Equation (3.5) above. For three vectors the equation becomes

$$\alpha_1 v_1 + \alpha_2 v_2 + \alpha_3 v_3 = 0 \qquad\qquad (3.6)$$

Substituting the actual values of the vectors in this equation and then multiplying by the scalars we obtain

$$[\alpha_1, \alpha_1, \alpha_1] + [\alpha_2, -2\alpha_2, \alpha_2] + [7\alpha_3, -11\alpha_3, +7\alpha_3] = [0,0,0]$$

This equation gives three equations once we have done the addition on the left-hand side and equated the coordinates of the resulting equal vectors.

The three equations are

$$\alpha_1 + \alpha_2 + 7\alpha_3 = 0$$
$$\alpha_1 - 2\alpha_2 - 11\alpha_3 = 0$$
$$a_1 + \alpha_2 + 7\alpha_3 = 0$$

We observe that we have two identical equations and hence we really only have two equations with three unknowns. From the first two equations we can obtain, by subtraction, the equation

$$3\alpha_2 + 18\alpha_3 = 0$$

Hence,

$$\alpha_2 = -6\alpha_3$$

Substituting this value for α_2 in the first equation we have the further equation

$$\alpha_1 = -\alpha_3$$

Hence we have a dependent set of vectors since the α_is are not necessarily all equal to zero.

The relationship (3.6) can now be rewritten by substituting α_1 and α_2 for their equivalent expressions in terms of α_3 to obtain the equation

$$-\alpha_3 v_1 - 6\alpha_3 v_2 + \alpha_3 v_3 = 0$$

which is equivalent to

$$- \alpha_3(v_1 + 6v_2 - v_3) = 0$$

The conclusion we can come to is that either α_3 equals zero or the expression in parentheses is equal to zero.

Check that the parentheses give the zero vector and thus α_3 could be any scalar. Hence, we have a relationship between the three original vectors given by the equation

$$v_1 + 6v_2 = v_3$$

which, by definition, shows dependency.

It is a good idea in practice to see if you can spot a relationship at the start!

The geometric interpretation of the equation of dependency arrived at above is that the vector v_3 must lie in the same plane as v_1 and v_2 because it is the sum of a linear combination of the vectors v_1 and v_2. We say that the vectors v_1, v_2, and v_3 are *coplanar*.

Exercise 3K

For each of the following sets of vectors, investigate whether they form an independent set or not:

(i) $v_1 = [0, 1, 2]$, $v_2 = [1, 3, 4]$, $v_3 = [-1, -1, 0]$
(ii) $v_1 = [1, 1, 1, 1]$, $v_2 = [0, 1, 1, 0]$, $v_3 = [-1, 2, -1, 4]$, $v_4 = [2, 4, 1, 7]$
(iii) $v_1 = [1, 0, 0, 0]$, $v_2 = [0, 1, 0, 0]$, $v_3 = [0, 0, 1, 0]$, $v_4 = [1, 1, 1, 1]$

DERIVE ACTIVITY 3E

(A)   Use DERIVE to check Example 3E and your solutions to Exercise 3K.
(B)   Consider a set of vectors which includes the zero vector. Will this set be dependent or independent?

Use DERIVE to answer this question by taking several examples of sets of vectors with and without the zero vector.

We will need the answer to this question later on in this book.

3.9   FURTHER VECTOR GEOMETRY

We now look at vector equations of lines and planes and show that they can be written in Cartesian form. We hope some of these forms will be familiar to you if you have done any coordinate geometry previously. First we will derive the equation of a line through a point with position vector **a** and parallel to a vector **b**.

Exercise 3L

Draw a triangle with vertices OAP. Mark $OA = $ **a**, $OP = $ **r** and $AP = \lambda$**b**, where λ is a scalar and **b** is a given vector.
        Show that the equation of the line through point A and parallel to **b** is

$$\mathbf{r} = \mathbf{a} + \lambda\mathbf{b}$$

In terms of Cartesian coordinates in Euclidean 3-space, we let $\mathbf{r} = [x, y, z]$, $\mathbf{a} = [a_1, a_2, a_3]$, and $\mathbf{b} = [b_1, b_2, b_3]$, then substituting these values into the equation $\mathbf{r} = \mathbf{a} + \lambda\mathbf{b}$ we obtain the three equations

$$x = a_1 + \lambda\, b_1; \quad y = a_2 + \lambda\, b_2; \quad z = a_3 + \lambda\, b_3$$

Now if none of the b_is are zero we can write three expressions all equal to λ.
        From the first equation $x - a_1 = \lambda\, b_1$ and thus $\lambda = \dfrac{x - a_1}{b_1}$. Hence,

$$\lambda = \frac{x - a_1}{b_1} = \frac{y - a_2}{b_2} = \frac{z - a_3}{b_3}$$

This is the Cartesian form of the vector equation of a straight line which passes through A and is parallel to the line joining $[0,0,0,]$ to $[b_1, b_2, b_3]$.

Example 3F

Find, in Cartesian form, the equation of the line through the point $[-2,0,5]$ and parallel to the vector $[1, 2, -3]$.

Solution

Direct substitution in the formula above gives:

$$\frac{x+2}{1} = \frac{y}{2} = \frac{z-5}{-3}$$

Example 3G

Find, in Cartesian form, the equation of the line through (1,2,3) and parallel to (–2,0,5).

Solution

Notice, first, that $b_2 = 0$, so we cannot use the formula found above directly. We need to go back to the original equations. Then

$$x = 1 - 2\lambda, \quad y = 2 + 0\lambda \quad \text{and} \quad z = 3 + 5\lambda$$

Hence we have $y = 2$ and $\dfrac{1-x}{2} = \dfrac{z-3}{5}$. This last equation can be written as $5x + 2z = 11$. The two equations $y = 2$ and $5x + 2z = 11$ define the required line in 3-space.

Exercise 3M

Show that the equation of a plane through the origin and parallel to two vectors **a** and **b** is $\mathbf{r} = \lambda\mathbf{a} + \mu\mathbf{b}$.

Hint. Proceed as before and let $\overrightarrow{OA} = \mathbf{a}$ and $\overrightarrow{OB} = \mathbf{b}$ with $\overrightarrow{OP} = \mathbf{r}$, where P will be any point on the line joining A and B.

Exercise 3N

Show that the equation of a plane through C and parallel to **a** and **b**, where C has coordinates $[c_1, c_2, c_3]$, is

$$\mathbf{r} = \mathbf{c} + \lambda\mathbf{a} + \mu\mathbf{b}$$

Example 3H

Find the equation of a plane in terms of a normal to the plane. (A normal to a plane is a vector which is perpendicular to or at right angles to that plane.)

Solution

Consider the plane through a given point A and perpendicular to a given vector $\mathbf{m}$. If $\mathbf{r}$ is the position vector of a general point P on the required plane and $\overrightarrow{OA} = \mathbf{a}$, then $\mathbf{r} - \mathbf{a}$ lies in the plane and is thus perpendicular to $\mathbf{m}$.
     Hence

$$(\mathbf{r} - \mathbf{a}) \cdot \mathbf{m} = 0 \qquad \text{(by the properties of the scalar or dot product)}$$

This is the equation of the plane since it is satisfied by any point on the plane but no points off the plane. In terms of Cartesian coordinates we have

$$\mathbf{r} = (x, y, z), \quad \mathbf{a} = (a_1, a_2, a_3) \quad \text{and} \quad \mathbf{m} = (m_1, m_2, m_3)$$

The equation $(\mathbf{r} - \mathbf{a}) \cdot \mathbf{m} = 0$ becomes

$$m_1(x - a_1) + m_2(y - a_2) + m_3(z - a_3) = 0$$

Now let $\mathbf{n}$ be a unit vector in the direction of $\mathbf{m}$ (directed from the origin to the plane). Then the above equation can be written

$$(\mathbf{r} - \mathbf{a}) \cdot \mathbf{n} = 0$$

This simplifies to

$$\mathbf{r} \cdot \mathbf{n} = \mathbf{a} \cdot \mathbf{n}$$

but $\mathbf{a} \cdot \mathbf{n}$ is the resolution of $\overrightarrow{OA}$ along $\mathbf{n}$ and is thus equal to the perpendicular from the origin to the plane. This distance is positive and is often denoted by p. Hence we have $(\mathbf{r} \cdot \mathbf{n}) = p$, which is known as the *normal form* of the equation of the plane.

Consolidation Exercise

1.   Consider the vectors **u** = [1, –3, 2] and **v** = [2, –1, 1] in $\mathfrak{R}^3$:

   (i)   Write down the vector [1, 7, – 4] as a linear combination of the vectors **u** and **v**.

   (ii)   For which values of k is the vector [1, k, 5] a linear combination of the vectors **u** and **v**.

   (iii)   Find a condition on a, b and c so that the vector [a b c] is a linear combination of the vectors **u** and **v**.

2.   Write the 2 × 2 matrix E as a linear combination of the matrices A, B and C where

$$A = \begin{bmatrix} 1 & 1 \\ 0 & -1 \end{bmatrix} \qquad B = \begin{bmatrix} 1 & 1 \\ -1 & 0 \end{bmatrix} \qquad C = \begin{bmatrix} 1 & -1 \\ 0 & 0 \end{bmatrix}$$

when

$$E = \begin{bmatrix} 3 & -1 \\ 1 & 2 \end{bmatrix}$$

3.   Using the same matrices for A, B and C in Problem 2, and taking E to be the matrix

$$\begin{bmatrix} 2 & 1 \\ -1 & -1 \end{bmatrix}$$

consider, with explanations, whether or not E can be written as a linear combination of A, B and C.

4.   Show that the vectors [1 1 1], [0 1 1], [0 1 –1] are a frame in the vector space $\mathfrak{R}^3$.

5.   Let **u**, **v** and **w** be vectors of the same size, and α any scalar.
        Using the definition of the scalar (or dot) product of two vectors, prove that the following properties hold in general:

   (i)   **u**·(**v** + **w**) = **u**·**v** + **u**·**w**

   (ii) α(**u**·**v**) = (α**u**)·**v** = **u**·(α**v**)

6.   Consider two vectors **u** and **v** in $\mathfrak{R}^2$, both of which are placed at the origin and with an angle θ, between them. Notice that θ could be acute or obtuse.
        The vector **u** can be projected on to the vector **v** (or vice versa). We define the projection of **u** on **v** as the vector

$$|\mathbf{u}| \cos(\theta) \left\{ \frac{\mathbf{v}}{|\mathbf{v}|} \right\}$$

Notice that the vector $\dfrac{\mathbf{v}}{|\mathbf{v}|}$ is another example of a unit vector. Its direction is the same as the direction of (**v**) but its length is 1.

The expression $|\mathbf{u}|\cos(\theta)$ is the resolution or perpendicular projection of the vector **u** on to the vector **v**. This idea of resolution is met in mechanics where forces are resolved in certain directions.

The problem for you to solve is to show that our definition is equivalent to the following definition, more usually found in textbooks.

The projection of the vector **u** on the vector **v** is given by

$$\frac{\mathbf{u}\cdot\mathbf{v}}{|\mathbf{v}|^2}\mathbf{v} = \frac{\mathbf{u}\cdot\mathbf{v}}{\mathbf{v}\cdot\mathbf{v}}\mathbf{v}$$

7.   Let **a**, **b** and **c** be three vectors forming a triangle. Label the angles opposite to these sides α, β and γ respectively.

Given that the vector product is distributive over vector addition, that is, equations of the following kind are true:

$$\mathbf{a}\times(\mathbf{a}+\mathbf{b}+\mathbf{c}) = \mathbf{a}\times\mathbf{a}+\mathbf{a}\times\mathbf{b}+\mathbf{a}\times\mathbf{c}$$

show that the sine rule

$$\frac{a}{\sin(\alpha)} = \frac{b}{\sin(\beta)} = \frac{c}{\sin(\gamma)}$$

follows.

Hint. The arrows on the vectors **a**, **b** and **c** should all be going in the same direction, that is, clockwise or anti-clockwise, and therefore the sum **a** + **b** + **c** must equal ... ??].

4

More Matrices

4.1   DEFINITIONS

Triangular and Diagonal Matrices

A square matrix A whose elements $a_{ij} = 0$ for $i > j$ is called *upper triangular*. A square matrix A whose elements $a_{ij} = 0$ for $i < j$ is called *lower triangular*.

The matrix A given below is upper triangular and the matrix B given below is lower triangular:

$$A = \begin{bmatrix} a_{11} & a_{12} & a_{13} & a_{14} & a_{15} \\ 0 & a_{22} & a_{23} & a_{24} & a_{25} \\ 0 & 0 & a_{33} & a_{34} & a_{35} \\ 0 & 0 & 0 & a_{44} & a_{45} \\ 0 & 0 & 0 & 0 & a_{55} \end{bmatrix} \qquad B = \begin{bmatrix} b_{11} & 0 & 0 \\ b_{21} & b_{22} & 0 \\ b_{31} & b_{32} & b_{33} \end{bmatrix}$$

A matrix D which is both upper and lower triangular is called a *diagonal matrix*. A 6×6 diagonal matrix D can be written in shorthand notation as

$$D = \text{diag}(d_{11}, d_{22}, d_{33}, d_{44}, d_{55}, d_{66})$$

that is we just list the diagonal elements, as all others are zero.

Notice that the identity or unit matrix introduced in Chapter 1 is another example of a diagonal matrix.

The 3 x 3 identity matrix could be written, diag (1,1,1).

The determinant of a triangular matrix is the product of the elements on the leading diagonal.

For the matrix B above, that is, the lower triangular matrix, we expand the determinant using the first row. Thus we have

$$\text{DET}(B) = b_{11} \begin{vmatrix} b_{22} & 0 \\ b_{32} & b_{33} \end{vmatrix} - 0 + 0$$

$$= b_{11}(b_{22}b_{33} - 0)$$

$$= b_{11}b_{22}b_{33}$$

Similarly for the matrix A above we would expand by the first column to obtain

$$\text{DET}(A) = a_{11}\, a_{22}\, a_{33}\, a_{44}\, a_{55}$$

As a diagonal matrix is both upper and lower triangular the determinant of such a matrix is also the product of the elements on the leading diagonal. In the case of matrix D above we would have

$$\text{DET}(D) = d_{11}\, d_{22}\, d_{33}\, d_{44}\, d_{55}\, d_{66}$$

In the case where there is a zero on the leading diagonal of a triangular or diagonal matrix, we can state at once that the determinant is zero.

Two square matrices A and B, of the same size, are said to be *commutative* if $AB = BA$. If A and B are such that $AB = -BA$, then the matrices A and B are said to be *anti-commutative* or we say A and B *anti-commute*.

Associativity

Three matrices A, B and C which are conformable under multiplication are said to be *associative under multiplication* if

$$(AB)C = A(BC)$$

In these definitions we assume that the sizes of the matrices are such that all the products exist. In general, we have seen that two matrices which are conformable under multiplication may not be commutative under multiplication.

DERIVE ACTIVITY 4A

(A)   Input the matrix $A = \begin{bmatrix} 1 & -1 \\ 2 & -1 \end{bmatrix}$ and the matrix $B = \begin{bmatrix} 1 & 1 \\ 4 & -1 \end{bmatrix}$.

Work out AB and BA by hand and check your results using DERIVE. You should find that A and B are anti-commutative.

(B)   Now consider $(A + B)^2$ for these two matrices.

Using DERIVE evaluate $(A + B)^2$ and also A^2 and B^2 separately. What relationship do you observe between these three matrices?

Generalize this result to any two 2×2 matrices C and D which you know to be anti-commutative?

(C)   Consider now the two general 2×2 matrices, $A = \begin{bmatrix} a & b \\ c & d \end{bmatrix}$ and $B = \begin{bmatrix} e & f \\ g & h \end{bmatrix}$ with

the property that A and B are anti-commutative.

Use DERIVE to find conditions on the entries so that numerical examples can be written down.

For example, by experiment and the use of algebra we find that

$$a = b = c = e = -d = -f = -g = -h$$

make A and B anti-commute.

By substituting these values into our general matrices we can check our

claim.   Check that if $A = \begin{bmatrix} a & a \\ a & -a \end{bmatrix}$ and $B = \begin{bmatrix} a & -a \\ -a & -a \end{bmatrix}$ then $AB = -BA$.

Hence we can write down lots of examples by replacing a by any number we

wish. For example, the matrices $\begin{bmatrix} 3 & 3 \\ 3 & -3 \end{bmatrix}$ and $\begin{bmatrix} 3 & -3 \\ -3 & -3 \end{bmatrix}$ are anti-commutative.

(D)   Try to find other conditions on the entries of the original general matrices so that more anti-commutative matrices can be written down.

Obtain conditions on the entries so that the two matrices we gave at the start of this activity can be generated?

Hint. By inspection of matrices A and B it looks like $a = e = f = -b = -d$ $= -h$ and $g = 2c$, but are these equations enough?

Periodic, Idempotent and Nilpotent Matrices

A matrix A for which $A^{k+1} = A$, where k is a positive integer, is called *periodic*. If k is the least positive integer for which this occurs then we say that A has period k. For the special case when $k = 1$ we have $A^2 = A$. We call a matrix with this property an *idempotent matrix*.

DERIVE ACTIVITY 4B

(A)   Input the matrix

$$A = \begin{bmatrix} 1 & -2 & -6 \\ -3 & 2 & 9 \\ 2 & 0 & -3 \end{bmatrix}$$

Use DERIVE to work out the period of A.
Is A idempotent?

(B)   Now consider the matrix

$$\begin{bmatrix} 0 & -\frac{1}{2} \\ 2 & -1 \end{bmatrix}$$

Use DERIVE again to find the period of this matrix.

(C)   Consider the general 2×2 matrix

$$\begin{bmatrix} a & b \\ c & d \end{bmatrix}$$

Try to find conditions on a, b, c and d so that the matrix will be periodic. Choose the period before you start.

Exercise 4A

1.   Show that the following matrices are periodic and find the period in each case. Which matrix is idempotent?

$$A = \begin{bmatrix} 1 & 0 \\ 4 & -1 \end{bmatrix} \qquad B = \begin{bmatrix} 2 & 4 \\ -\frac{1}{2} & 1 \end{bmatrix}$$

(Clearly this activity lends itself to DERIVE.)

2.   Find a condition on a and b for the diagonal matrix

$$D = \begin{bmatrix} a & 0 \\ 0 & b \end{bmatrix}$$

to be periodic of period k.

3.   Find a condition on c and d for the matrix

$$E = \begin{bmatrix} 0 & c \\ d & 0 \end{bmatrix}$$

to be periodic. Can E be idempotent?

Nilpotent Matrix

A matrix A for which $A^P = 0$ (the zero matrix), where p is a positive integer, is said to be

nilpotent

If p is the least positive integer for which this occurs we say that A is

nilpotent of index p

For example the matrix $A = \begin{bmatrix} 0 & 0 \\ 1 & 0 \end{bmatrix}$ is nilpotent of index 2 because

$$\begin{bmatrix} 0 & 0 \\ 1 & 0 \end{bmatrix} \begin{bmatrix} 0 & 0 \\ 1 & 0 \end{bmatrix} = \begin{bmatrix} 0 & 0 \\ 0 & 0 \end{bmatrix}$$

DERIVE ACTIVITY 4C

(A)   Consider the matrix $A = \begin{bmatrix} 0 & 1 \\ 0 & 0 \end{bmatrix}$.

Work out A^2, A^3, etc., until $A^p = 0$, the zero matrix. Check your calculations using DERIVE. You should find that A is nilpotent of index 2.

Notice that, because we started to work out the powers of A from the bottom, that is, squared then cubed, etc., we arrive at the index p, the least positive integer.

(B)   Using DERIVE show that the following matrices are nilpotent and work out the index for each:

$$B = \begin{bmatrix} 1 & 1 & 3 \\ 5 & 2 & 6 \\ -2 & -1 & -3 \end{bmatrix} \qquad C = \begin{bmatrix} 1 & -3 & -4 \\ -1 & 3 & 4 \\ 1 & -3 & -4 \end{bmatrix}$$

Use DERIVE to experiment with other matrices to see whether or not they are nilpotent.

Can you find general matrices 2×2, and 3×3 which will be nilpotent?

4.2   THE MULTIPLICATIVE INVERSE REVISITED!

In Chapter 1, we introduced the idea of the multiplicative inverse of a matrix A which we labelled A^{-1}, in keeping with multiplicative notation.

We now want to extend our knowledge. Let A and B be two square non-singular matrices, that is, matrices which have inverses. A question we might ask is: do the products AB and BA have multiplicative inverses, and if so how are they related to A^{-1} and B^{-1}? The answer is that $(AB)^{-1} = B^{-1}A^{-1}$ and $(BA)^{-1} = A^{-1}B^{-1}$.

Proof

We assume that both A and B have inverses as stated above. Consider the product matrix AB and assume it has an inverse. Then, by definition,

$$(AB)(AB)^{-1} = I$$

Now because matrices in general do not commute under multiplication we have to be careful how we manipulate this equation. We multiply both sides of this equation **on the left** by A^{-1} to obtain

$$A^{-1}(AB)(AB)^{-1} = A^{-1}I$$

By associativity of matrices and the fact that $AA^{-1} = I$, we can simplify this equation to

$$B(AB)^{-1} = A^{-1}$$

Now multiply both sides of this equation by B^{-1} **on the left** to obtain

$$B^{-1}B(AB)^{-1} = B^{-1}A^{-1}$$

Again remembering that $B^{-1}B = I$ and $CI = IC = C$ for any C conformable with the identity matrix I, the above equation simplifies to the result

$$(AB)^{-1} = B^{-1}A^{-1}$$

In a similar way we can show that $(BA)^{-1} = A^{-1}B^{-1}$.

By a similar argument we can prove the result for n matrices:

$$(A_1 A_2 A_3 A_4 A_5 \ldots A_n)^{-1} = A_n^{-1} \ldots A_5^{-1} A_4^{-1} A_3^{-1} A_2^{-1} A_1^{-1}$$

Involutary Matrices

A matrix A for which $A^2 = I$ is called *involutary*. It follows that such a matrix is its own inverse because

$$A \cdot A^{-1} = I = A^{-1} \cdot A$$

Furthermore, the identity matrix I is an example of such a matrix.

DERIVE ACTIVITY 4D

(A)   The matrix $\begin{bmatrix} 0 & 1 \\ 1 & 0 \end{bmatrix}$ is involutary.

   Check by hand and also by using DERIVE.

(B)   Working by hand and checking using DERIVE, ascertain which, if any, of the following matrices are involutary?

$$A = \begin{bmatrix} 0 & 1 & -1 \\ 4 & -3 & 4 \\ 3 & -3 & 4 \end{bmatrix} \qquad B = \begin{bmatrix} 1 & 2 & -3 \\ 0 & 1 & 1 \\ 4 & 3 & -4 \end{bmatrix} \qquad C = \begin{bmatrix} 4 & 3 & 3 \\ -1 & 0 & -1 \\ -4 & -4 & -3 \end{bmatrix}$$

4.3   THE ADJOINT OR ADJUGATE MATRIX

In this section we again consider a square matrix A and we need to recall the idea of cofactors from our earlier work. In what follows, we will use α_{ij} for the cofactor of a_{ij}. We now define the adjugate or *adjoint matrix* of A, denoted adj(A).

Definition

If $A = (a_{ij})$ then adj$(A) = (\alpha_{ij})^T$. This means that in the original matrix A we replace every one of its elements by its appropriate cofactor and then transpose the resulting matrix to obtain the adjugate or adjoint of A.

DERIVE ACTIVITY 4E

(A)   Consider again the matrix $A = \begin{bmatrix} 1 & 2 \\ 3 & 4 \end{bmatrix}$

The cofactor of element $a_{11} = 1$ will, by definition, be $(-1)^{1+1} |4| = 4$.

Work out the other three cofactors in the same way. (Again use DERIVE to verify your results and to give you practice in using the cofactor function introduced in Chapter 2.)

Now replace each of the four elements in A by their cofactors to obtain the matrix

$$\begin{bmatrix} 4 & -3 \\ -2 & 1 \end{bmatrix}$$

Then, by definition,

$$\text{adj}(A) = \begin{bmatrix} 4 & -3 \\ -2 & 1 \end{bmatrix}^T = \begin{bmatrix} 4 & -2 \\ -3 & 1 \end{bmatrix}$$

(B)   Now consider a 3×3 matrix B:

$$\text{Verify that } \text{adj}(B) = \begin{bmatrix} 6 & 1 & -5 \\ -2 & -5 & 4 \\ -3 & 3 & -1 \end{bmatrix} \text{ when } B = \begin{bmatrix} 1 & 2 & 3 \\ 2 & 3 & 2 \\ 3 & 3 & 4 \end{bmatrix}$$

Again use DERIVE to verify results obtained by hand.

Before we look at the properties of the adjoint matrix we digress to introduce the idea of *alien cofactors*. These are not cofactors from another planet, but, as we will see, they will be extremely useful to us in obtaining properties of adjoints.

Let $A = (a_{ij})$ be a general $n \times n$ matrix and let $\text{adj}(A) = (\alpha_{ij})^T$, as described at the beginning of this section.

Now consider the matrix A with its jth column replaced by

$$k_1, k_2, k_3, k_4, ..., k_n$$

that is, a_{1j} is replaced by k_1, and a_{2j} by k_2, and so on.

Call this new matrix B.

Now expand DET(B) using this column. We can write this expansion as follows:

$$k_1\alpha_{1j} + k_2\alpha_{2j} + ... + k_n\alpha_{nj} = \text{DET(B)}$$

Now if the k_is are exactly one of the **other** columns in the matrix A or B, then we claim that DET(B) = 0.

Assume that this claim is true, that is, two columns in DET(B) are now identical. We can subtract these two columns from each other in evaluating the determinant and will then obtain a column of zeros.

Hence using the first property of determinants given in Chapter 2 we obtain the value of zero for DET(B), as claimed. For example, if the k_is column is the same as column 1 we have just shown that

$$a_{11}\alpha_{1j} + a_{21}\alpha_{2j} + ... + a_{n1}\alpha_{nj} = 0$$

Similar expressions can be written down using any other column (except the jth), for the k_is column.

Hence what we have shown is that if we expand a determinant using any column and using the cofactors from any **other** (alien) column the result must be zero. We then say that we have expanded the determinant by *alien cofactors*.

As an example of this idea consider the determinant C given below:

$$C = \begin{vmatrix} 1 & 2 & 3 \\ 0 & 1 & 2 \\ 1 & 0 & 2 \end{vmatrix}$$

Now expand the determinant using the first column with the cofactors of the third column. The calculation is

$$\mathrm{DET}(C) = 1 \times \begin{vmatrix} 0 & 1 \\ 1 & 0 \end{vmatrix} - 0 \times \begin{vmatrix} 1 & 2 \\ 1 & 0 \end{vmatrix} + 1 \times \begin{vmatrix} 1 & 2 \\ 0 & 1 \end{vmatrix} = -1 - 0 + 1 = 0$$

Exercise 4B

Expand the determinant C using the first column again with the cofactors of the second column. The result should still be zero. Only when we use the first column with its own cofactors will we obtain the true value of $\mathrm{DET}(C)$.

Now, after that digression, we can return to look at properties of the adjoint matrix.

Properties of the Adjoint or Adjugate Matrix

The adjoint matrix has some interesting properties and is most useful in applications, as we shall see.

Let $A = (a_{ij})$ be a general $n \times n$ matrix and let $\mathrm{adj}(A) = (\alpha_{ij})^T$.

Result 1

$$A(\mathrm{adj}(A)) = \mathrm{diag}(|A|, |A|, \dots, |A|)$$
$$= |A| \cdot I_n$$
$$= (\mathrm{adj}(A))A$$

This result shows that A and its adjoint commute. To prove this result we will make good use of our alien cofactor work. Start by writing out the general matrices for A and $\mathrm{adj}(A)$ as given below:

$$A(\text{adj}(A)) = \begin{bmatrix} a_{11} & a_{12} & a_{13} & \cdots \\ a_{21} & a_{22} & a_{23} & \cdots \\ \vdots & \vdots & \vdots & \end{bmatrix} \cdot \begin{bmatrix} \alpha_{11} & \alpha_{21} & \alpha_{31} & \cdots \\ \alpha_{12} & \alpha_{22} & \alpha_{32} & \cdots \\ \vdots & \vdots & \vdots & \end{bmatrix}$$

Remember that the adjoint matrix is the cofactor matrix transposed, hence the labelling of the αs. Now work out this product to obtain entries in the top left-hand corner as,

$$a_{11}\alpha_{11} + a_{12}\alpha_{12} + a_{13}\alpha_{13} + \ldots \qquad a_{11}\alpha_{21} + a_{12}\alpha_{22} + a_{13}\alpha_{23} + \ldots$$

$$a_{21}\alpha_{11} + a_{22}\alpha_{12} + a_{23}\alpha_{13} + \ldots \qquad a_{21}\alpha_{21} + a_{22}\alpha_{22} + a_{23}\alpha_{23} + \ldots$$

$$a_{31}\alpha_{11} + a_{32}\alpha_{12} + a_{33}\alpha_{13} + \ldots$$

From these few entries we see that entries on the leading diagonal will all give DET(A) and those off the leading diagonal will all be zero because they are alien cofactors for the a_{ij}s that are multiplying them. For example,

$$a_{11}\alpha_{21} + a_{12}\alpha_{22} + a_{13}\alpha_{23} + \ldots = 0$$

because the first row of A is being used with the cofactors of the second row. Thus we have proved that

$$A(\text{adj}(A)) = \text{diag}(|A|, |A|, \ldots, |A|)$$

In a similar way,

$$\text{adj}(A))A = \text{diag}(|A|, |A|, \ldots, |A|)$$

Finally, the diagonal matrix can be factorized to give

$$|A| \cdot I_n$$

where I_n is the $n \times n$ identity or unit matrix. Thus Result 1 has been proved.

Result 2

$$|A||\text{adj}(A)| = |A|^n$$

This says that the determinant of A multiplied by the determinant of its adjoint is equal to the determinant of A raised to the nth power.

Proof

This is obtained from Result 1 by taking determinants of the equation

$$A(\mathrm{adj}(A)) = \mathrm{diag}(|A|, |A|, \dots, |A|)$$

Remember that $\mathrm{DET}(C \cdot D) = \mathrm{DET}(C) \cdot \mathrm{DET}(D)$ from Chapter 2, and this is proved later in this chapter.

Also remember that the determinant of a diagonal matrix will simply be the elements on the leading diagonal multiplied together.

Using these results on the above equation we have

$$\mathrm{DET}(A) \cdot \mathrm{DET}(\mathrm{adj}(A)) = (\mathrm{DET}(A))^n$$

This is Result 2 written using DET instead of $|\ |$.

From Result 2, it follows that if $\mathrm{DET}(A) \neq 0$, we can divide both sides of the equation by $\mathrm{DET}(A)$, and obtain

$$|\mathrm{adj}(A)| = |A|^{n-1}$$

However, if $\mathrm{DET}(A) = 0$ then $A(\mathrm{adj}(A)) = 0$ (the zero matrix), using Result 1.

Exercise 4C

Using any results obtained so far prove the following:

(i) $\mathrm{adj}(AB) = \mathrm{adj}(B)\mathrm{adj}(A)$.

(ii) $\mathrm{adj}(\mathrm{adj}(A)) = |A|^{n-2}A$ if $\mathrm{DET}(A) \neq 0$.

Now we can write the inverse of a matrix in another form. From the result

$$A(\mathrm{adj}(A)) = |A|I_n$$

and assuming A has an inverse A^{-1} we obtain the equation

$$A^{-1}A(\mathrm{adj}(A)) = A^{-1}|A|I_n$$

It follows that if $\mathrm{DET}(A) = |A| \neq 0$, then:

$$A^{-1} = \frac{\mathrm{adj}(A)}{|A|}$$

This definition enables us to obtain inverses from the adjoint or adjugate matrix.

Example 4A

Use the adjoint matrix to find the inverse of the matrix

$$A = \begin{bmatrix} 1 & 2 \\ 3 & 4 \end{bmatrix}$$

Solution

The adjoint matrix of A is

$$\text{adj}(A) = \begin{bmatrix} 4 & -3 \\ -2 & 1 \end{bmatrix}^T = \begin{bmatrix} 4 & -2 \\ -3 & 1 \end{bmatrix}$$

$$\text{DET}(A) = \begin{vmatrix} 1 & -2 \\ -3 & 4 \end{vmatrix} = -2$$

Hence,

$$A^{-1} = \frac{\text{adj}(A)}{\text{DET}(A)} = -\frac{1}{2}\begin{bmatrix} 4 & -2 \\ -3 & 1 \end{bmatrix} = \begin{bmatrix} -2 & 1 \\ 1.5 & -0.5 \end{bmatrix}$$

We can check that this is the inverse because

$$\mathbf{AA}^{-1} = \begin{bmatrix} 1 & 2 \\ 3 & 4 \end{bmatrix} \cdot \begin{bmatrix} -2 & 1 \\ 1.5 & -0.5 \end{bmatrix} = \begin{bmatrix} -2+3 & 1-1 \\ -6+6 & 3-2 \end{bmatrix} = \begin{bmatrix} 1 & 0 \\ 0 & 1 \end{bmatrix}$$

Exercise 4D

Use the adjoint matrix method to find the inverse of the following matrices:

$$C = \begin{bmatrix} -1 & -3 \\ 1 & 2 \end{bmatrix} \qquad B = \begin{bmatrix} 1 & 2 & 3 \\ 2 & 3 & 2 \\ 3 & 3 & 4 \end{bmatrix}$$

4.4   STOCHASTIC MATRICES

Definitions

A matrix A is said to be *stochastic* or (*row-stochastic*) if

> each of its entries is non-negative, and
> each of its rows sum to 1.

In some books this type of matrix is called a *probability matrix*.
Hence if a matrix is such that

> each of its entries is non-negative, and
> each of its columns sum to 1,

its transpose is (row) stochastic.
If a matrix with non-negative entries has

> each of its rows, and
> each of its columns summing to 1,

then it is said to be *doubly stochastic*. These matrices play a leading role in statistics, not least in stochastic processes.
We will endeavour to give some simple examples in Chapter 9.

4.5   ECHELON MATRICES

Definition

An echelon matrix is a matrix with the following two properties:

(i)   The zero rows, if any, occur at the bottom of the matrix, that is, the bottom rows of the matrix.

(ii)   If $R_1, R_2, R_3, ..., R_k$ are the non-zero rows (in order from the top), then the first non-zero entry of R_{i+1} occurs strictly further right than the first non-zero entry of R_i.

> As an example consider the matrix A given below. It has its zero rows at the bottom of the matrix, row 4 and row 5 being zero rows. Furthermore, looking at row 1, row 2 and row 3 in order, we see that the first non-zero entry of row 2 is strictly further to the right than in row 1. Also, the first non-zero entry of row 3 occurs strictly further to the right than the first non-zero entry of row 2. Therefore, the matrix A is an echelon matrix:

$$A = \begin{bmatrix} 1 & -2 & 3 & -1 \\ 0 & 1 & 2 & 1 \\ 0 & 0 & 1 & 2 \\ 0 & 0 & 0 & 0 \\ 0 & 0 & 0 & 0 \end{bmatrix}$$

A *reduced echelon matrix* satisfies the two properties given above as well as the two further properties given below:

(iii)   In each non-zero row the leading entry (the first non-zero entry) is always 1.

(iv)   In each column that contains the leading entry of a row, all other entries are zero.

The matrix A given above is **not** a reduced echelon matrix, because even though it satisfies Property (iii) it does not satisfy Property (iv). The matrix B given below is a reduced echelon matrix, as can be seen by checking that each of the four properties is satisfied:

$$B = \begin{bmatrix} 1 & 0 & 2 & 0 & -4 \\ 0 & 1 & 7 & 0 & 0 \\ 0 & 0 & 0 & 1 & 3 \\ 0 & 0 & 0 & 0 & 0 \end{bmatrix}$$

Distinguished Elements

The elements whose values are 1 in the columns where the rest of the entries are zero are sometimes referred to as *distinguished elements*. Thus there are three distinguished elements in matrix B above:

b_{11}, b_{22} and b_{34}.

4.6   ELEMENTARY ROW OPERATIONS AND ROW EQUIVALENCE

Given a matrix A we can perform elementary row operations on it to obtain another matrix B, which we will say is related to matrix A. This idea will be formalized a little later in this section.

The *elementary row operations* are

(i)   interchange the ith row and the jth row: $R_i \leftrightarrow R_j$
(ii)   multiply the ith row by a non-zero scalar k: $R_i \rightarrow kR_i$
(iii)   replace the ith row by k times the jth row plus the ith row:

$$R_i \rightarrow kR_j + R_i$$

As an example consider the matrix A given below:

$$A = \begin{bmatrix} 1 & 2 & 3 \\ 2 & 1 & 3 \\ 3 & 1 & 2 \end{bmatrix}$$

Consider now the first operation, that of interchanging two rows.

Let us interchange row 3 and row 1. This instruction could be written $R_1 \leftrightarrow R_3$.

We obtain the matrix

$$\begin{bmatrix} 3 & 1 & 2 \\ 2 & 1 & 3 \\ 1 & 2 & 3 \end{bmatrix}$$

As an example of row operation (ii) let us multiply row 2 of the matrix A by the scalar $-\frac{1}{2}$. This instruction could be written $R_2 \rightarrow (-\frac{1}{2})R_2$.

We obtain the matrix

$$\begin{bmatrix} 1 & 2 & 3 \\ -1 & -0.5 & -1.5 \\ 3 & 1 & 2 \end{bmatrix}$$

Finally, let us demonstrate Property (iii) by adding –2 lots of row 1 to row 2, to replace row 2. This instruction could be written $R_2 \rightarrow (-2)R_1 + R_2$.

We obtain the matrix

$$\begin{bmatrix} 1 & 2 & 3 \\ 0 & -3 & -3 \\ 3 & 1 & 2 \end{bmatrix}$$

We commented earlier that the matrix A and the matrix obtained from A by performing an elementary operation on it would be related. We now formalize this idea.

Definition

A matrix A is said to be *row equivalent* to a matrix B if B can be obtained from A by a finite sequence of elementary row operations. The terminology *row reduce* means to transform by elementary row operations.

Exercise 4E

For each of the matrices B and C, find the row equivalent matrices according to the following elementary row operations:

(i) $R_2 \leftrightarrow R_4$

(ii) $R_3 \rightarrow -3R_1 + R_3$

$$B = \begin{bmatrix} 0 & 1 & 2 & 1 \\ 0 & 3 & -1 & 4 \\ 2 & 1 & 0 & -3 \\ 1 & -1 & 2 & 1 \end{bmatrix} \qquad C = \begin{bmatrix} 1 & 0 & -1 & 1 \\ 2 & -1 & 3 & -4 \\ 0 & 0 & 1 & 2 \\ 5 & -1 & 2 & 0 \end{bmatrix}$$

4.7   ELEMENTARY MATRICES

Definition

A matrix obtained from the identity matrix I, by a single elementary row operation is called an *elementary matrix*.

Example 4B

Starting with the 3 × 3 identity matrix I, perform the following elementary row operations to obtain new matrices which are row equivalent to I and label them as three elementary matrices E_1, E_2, and E_3:

(i) $R_1 \rightarrow 10R_1$
(ii) $R_2 \rightarrow R_2 + 3R_3$
(iii) $R_1 \leftrightarrow R_3$

Solutions

(i) $R_1 \rightarrow 10R_1$ is an elementary operation so $I \rightarrow \begin{bmatrix} 10 & 0 & 0 \\ 0 & 1 & 0 \\ 0 & 0 & 1 \end{bmatrix} = E_1$

(ii) $R_2 \rightarrow R_2 + 3R_3$ is an elementary operation so $I \rightarrow \begin{bmatrix} 1 & 0 & 0 \\ 0 & 1 & 3 \\ 0 & 0 & 1 \end{bmatrix} = E_2$

(iii) $R_1 \leftrightarrow R_3$ is an elementary row operation so $I \rightarrow \begin{bmatrix} 0 & 0 & 1 \\ 0 & 1 & 0 \\ 1 & 0 & 0 \end{bmatrix} = E_3$

Remember, E_1, E_2 and E_3 are examples of elementary matrices.
    Notice that these are the three basic row operations and, furthermore, there is something special about their matrices:

$$\text{DET}(E_1) = 10 \qquad \text{DET}(E_2) = 1 \qquad \text{DET}(E_3) = -1$$

We will generalize these remarks later in this chapter.
    Using the ideas given above about elementary matrices, we now have yet another way to compute the inverse of a matrix, when it exists. Let us assume that we know that the inverse of a square matrix A exists. The trick is that we perform elementary row operations on A in order eventually to arrive at the identity matrix, I.

Meanwhile, we perform exactly the same operations, in the same order, on the identity matrix, I, and the identity matrix transforms into the required inverse!

Augmented Matrix

To carry out the above procedure, we use the idea of an augmented matrix. Let A be a 3×3 matrix. An example of an augmented matrix is $[A|\,I]$, where I is the 3×3 identity matrix. Here we have placed the matrix A side-by-side with the identity matrix I. We have augmented the matrix A by (or with)the identity matrix I to obtain a 3×6 matrix.

Often, we want to keep the two original matrices separate so we place a vertical bar between them. In this way we can operate on both matrices at the same time and keep track of what is happening.

Example 4C

Consider the matrix

$$A = \begin{bmatrix} 2 & 1 & -2 \\ 3 & 2 & 2 \\ 5 & 4 & 3 \end{bmatrix}$$

Augment A using the 3×3 identity matrix and find the inverse of A using elementary row operations.

Solution

The augmented matrix is

$$[A|I] = \begin{bmatrix} 2 & 1 & -2 & 1 & 0 & 0 \\ 3 & 2 & 2 & 0 & 1 & 0 \\ 5 & 4 & 3 & 0 & 0 & 1 \end{bmatrix}$$

We are now going to perform elementary row operations on this augmented matrix to obtain the augmented matrix given below in nine steps:

$$\begin{bmatrix} 1 & 0 & 0 & 2/7 & 11/7 & -6/7 \\ 0 & 1 & 0 & -1/7 & -16/7 & 10/7 \\ 0 & 0 & 1 & -2/7 & 3/7 & -1/7 \end{bmatrix}$$

The first step would be to divide the first row of $[A|I]$ by 2. We obtain

$$\begin{bmatrix} 1 & 0.5 & -1 & 0.5 & 0 & 0 \\ 3 & 2 & 2 & 0 & 1 & 0 \\ 5 & 4 & 3 & 0 & 0 & 1 \end{bmatrix}$$

This can be represented by an elementary matrix E_1:

$$E_1 = \begin{bmatrix} 0.5 & 0 & 0 \\ 0 & 1 & 0 \\ 0 & 0 & 1 \end{bmatrix}$$

and the product $E_1 A$ will be the matrix with the first row divided by 2. So the first step has transformed $[A \,|\, I]$ to $[E_1 A \,|\, E_1]$.

Our next step is to use the first row of this matrix to eliminate the 3 in position a_{21}. We can do this by taking three lots of row 1 away from row 2. In symbols:

$$R_2 \rightarrow R_2 - 3R_1$$

The elementary matrix

$$E_2 = \begin{bmatrix} 1 & 0 & 0 \\ -3 & 1 & 0 \\ 0 & 0 & 1 \end{bmatrix}$$

has the same effect.

Computing $E_2 E_1 A$ and $E_2 E_1 I$ we have now transformed $[A|I]$ to $[E_2 E_1 A | E_2 E_1 I]$. Written in full we have

$$\begin{bmatrix} 1 & 0.5 & -1 & 0.5 & 0 & 0 \\ 0 & 0.5 & 5 & -1.5 & 1 & 0 \\ 5 & 4 & 3 & 0 & 0 & 1 \end{bmatrix}$$

Our next step would be to eliminate the 5 in position a_{31} by taking five lots of row 1 away from row 3. The reader is left to perform this row operation and to write it in symbols and to arrive at the matrix

$$[E_3 E_2 E_1 A | E_3 E_2 E_1 I]$$

where E_3 is the elementary matrix obtained from the identity matrix by performing the row operation just described above. We give E_3 below to ensure that you have followed every step so far:

$$E_3 = \begin{bmatrix} 1 & 0 & 0 \\ 0 & 1 & 0 \\ -5 & 0 & 1 \end{bmatrix} \quad \text{and} \quad [E_3 E_2 E_1 A | E_3 E_2 E_1 I] = \begin{bmatrix} 1 & 0.5 & -1 & 0.5 & 0 & 0 \\ 0 & 0.5 & 5 & -1.5 & 1 & 0 \\ 0 & 1.5 & 8 & -2.5 & 0 & 1 \end{bmatrix}$$

At this stage we concentrate on the element $a_{22} = 0.5$. We multiply row 2 by 2 in order to make this element 1. This will involve an elementary matrix, E_4.

Then we use the second row to make the elements a_{12} and a_{32} zero. This will involve two further elementary matrices, E_5 and E_6.

Finally, we make the element a_{33} equal to 1 by another elementary matrix E_7 and then make the remaining two elements a_{31} and a_{32} equal to zero. This will involve the last two elementary matrices E_8 and E_9.

If you have completed each of these stages correctly you should have arrived at the following 3×6 matrix:

$$\begin{bmatrix} 1 & 0 & 0 & 2/7 & 11/7 & -6/7 \\ 0 & 1 & 0 & -1/7 & -16/7 & 10/7 \\ 0 & 0 & 1 & -2/7 & 3/7 & -1/7 \end{bmatrix}$$

Check, using DERIVE or by hand, that the 3×3 matrix to the right of the identity matrix is indeed the inverse of the original matrix we started with.

Let us now summarize this procedure in theory.

You will have observed, hopefully, that for a 3×3 matrix a maximum of nine elementary row operations sufficed to achieve our aim. It could be that less than nine would be needed in some cases, for example if an entry of 1 was already in position a_{11}, we would not need to divide the first row by a_{11} as we did in the example given above.

The theory given below extends to any square matrix. If the original matrix were $n \times n$ then we would need at most n^2 elementary matrices.

We start with the equation

$$E_9 E_8 E_7 E_6 E_5 E_4 E_3 E_2 E_1 A = I \qquad (4.1)$$

This shows the nine elementary matrices operating on the original matrix A and the result is the identity matrix I. What we want to show is that if the same nine elementary matrices operate in the same order, on the identity matrix I, we obtain A^{-1}, the inverse of our original matrix A. This always assumes that the inverse exists.

Now let us assume that A^{-1} exists. Now multiply both sides of Equation (4.1) on the right by A^{-1}. We now have the following equation:

$$E_9 E_8 E_7 E_6 E_5 E_4 E_3 E_2 E_1 A A^{-1} = I A^{-1} = A^{-1}$$

Recall that $I \cdot B = B$ and $B^{-1} \cdot B = I$ for any matrix B. Hence, the equation simplifies to

$$E_9 E_8 E_7 E_6 E_5 E_4 E_3 E_2 E_1 I = A^{-1}$$

This final equation shows that if we perform the same nine elementary operations (given by elementary matrices) on I, we will indeed obtain the inverse of A.

Similarly, if we begin with

$$E_9 E_8 \dots E_2 E_1 A^{-1} = I$$

we can multiply on the right of each side of the above equation to obtain

$$E_9 E_8 \dots E_2 E_1 A^{-1} A = A$$

which reduces to

$$E_9 E_8 \dots E_2 E_1 I = A$$

This shows that any matrix A which has an inverse can be written as the product of elementary matrices. This fact holds for any $n \times n$ invertible matrix.

Exercise 4F

Using elementary row operations given by elementary matrices find the inverse of the following matrices:

$$A = \begin{bmatrix} 1 & -1 \\ 2 & 3 \end{bmatrix} \qquad B = \begin{bmatrix} 2 & 0 & 1 \\ -1 & 2 & 3 \\ -3 & -1 & 1 \end{bmatrix}$$

For each matrix list the elementary matrices.

Determinants of Elementary Matrices

In this section we show that the determinant of an elementary matrix is either +1, −1 or k, where k will be defined below. Recall Example 4B earlier in this chapter. There we had an example of each type of elementary matrix, obtained from the identity matrix by performing each of the elementary row operations defined in Section 4.7. We observed that the determinant was either +1, −1 or 10, this latter value coming from the fact that we had multiplied row 1 by a factor of 10. This will generalize to k in what follows. To prove the statement at the beginning of this section we examine the three types of elementary matrix and then compute their determinants.

1.   *Interchange of the* ith *and* jth *rows.*

We have seen previously that det(I) = 1. Here we start with I and interchange the ith and jth row to obtain an elementary matrix E. By Property (iii) of determinants, given in Chapter 2,

DET(E) = −DET(I)

and hence,

DET(E) = −1

2.   *Multiply the* ith *row by a non-zero scalar* k.

Again DET(I) = 1 and the ith row of I has been multiplied by k to obtain an elementary matrix E. By Property (iv) of determinants, given in Chapter 2, we can factorize E, since the ith row now has a factor of k in each of its entries. Hence we have

DET(E) = k·DET(I)

Hence,

$$\text{DET}(E) = k$$

3.   *Replace the ith row by k times the jth row plus the ith row.*

Again $\text{DET}(I) = 1$ and we start with the identity matrix and perform this elementary operation on it to obtain the elementary matrix E. Here we appeal to Property (v) of determinants given in Chapter 2. Hence,

$$\text{DET}(E) = \text{DET}(I)$$

that is,

$$\text{DET}(E) = 1$$

Next we use these results to prove the following fact:

$$\text{DET}(E\cdot A) = \text{DET}(E)\cdot\text{DET}(A)$$

where E is an elementary $n \times n$ matrix and A is any $n \times n$ matrix.
         We prove this by considering again the three possibilities for E and the effect that E will have on A in the multiplication, $E\cdot A$, and the value of the determinants on both sides of the equation given above.

1.   *Let E be the operation $R_i \leftrightarrow R_j$ and A an $n \times n$ matrix.*
         Consider first $E.A$, which will be the matrix A with two rows interchanged. Hence,

$$\text{DET}(E\cdot A) = -\text{DET}(A)$$

by Property (iii) in Chapter 2.
         Now consider the product, $\text{DET}(E)\cdot\text{DET}(A)$. We have seen above, that $\text{DET}(E)$ in this case is -1; hence,

$$\text{DET}(E)\cdot\text{DET}(A) = -1\cdot\text{DET}(A) = -\text{DET}(A)$$

Thus $\text{DET}(E\cdot A) = \text{DET}(E)\cdot\text{DET}(A)$, as required.

2.   *Let E be the operation $R_i \leftrightarrow R_i + kR_j$ and A an $n \times n$ matrix.*
         Again consider the product $E\cdot A$, which will be the matrix A changed by replacing row i by row i added to k lots of row j. By Property (v) given in Chapter 2 we then have

$$DET(E \cdot A) = DET(A)$$

Furthermore $DET(E)$ in this case has the value $+1$ from above and thus

$$DET(E) \cdot DET(A) = +1DET(A) = DET\ A$$

Hence we have in this case

$$DET(E \cdot A) = DET(E) \cdot DET(A)$$

3. *Let E be the operation* $R_i \leftrightarrow kR_i$ *and let A be an* $n \times n$ *matrix.*
    The product $E \cdot A$ in this case will be the matrix A with its ith row multiplied throughout by k. By Property (iv) given in Chapter 2 it follows that

$$DET(E \cdot A) = k \cdot DET(A)$$

However, in this case we have shown earlier that $DET(E) = k$ and hence

$$DET(E \cdot A) = DET(E) \cdot DET(A)$$

as required.

With this result established we can proceed to prove, in part, the important fact about determinants used earlier in this book and given below.
Let A and B be any two $n \times n$ matrices; then

$$DET(A \cdot B) = DET(A) \cdot DET(B)$$

We have three cases to consider. We will consider one here. The remaining two cases will be completed in the next chapter when we have established a little more theory. The proof will be more straightforward at that stage than it would be now.
The three cases are

(i) $DET(A) \neq 0$ and $DET(B) \neq 0$.
(ii) $DET(A) = 0$ and $DET(B) \neq 0$, or vice versa.
(iii) Both $DET(A)$ and $DET(B)$ are zero.

Proof of (i)

Since $DET(A)$ and $DET(B)$ are non-zero, by our earlier work, A^{-1} and B^{-1} exist. Also from earlier work, since A^{-1} exists it can be written as a product of elementary matrices and so therefore can A as follows:

$$E_n E_{n-1} \ldots E_2 E_1 = A \qquad (4.2)$$

Therefore, if we multiply both sides of this equation on the right by B, we obtain

$$E_n E_{n-1} \ldots E_2 E_1 B = AB$$

Now take determinants of both sides to obtain

$$\text{DET}(E_n E_{n-1} \ldots E_2 E_1 B) = \text{DET}(AB)$$

Using the fact proved earlier in this chapter that $\text{DET}(EB) = \text{DET}(E)\text{DET}(B)$ we have

$$\text{DET}(E_n)\text{DET}(E_{n-1} \ldots E_2 E_1 B) = \text{DET}(AB)$$

Applying this fact repeatedly, we arrive at the equivalent equation:

$$\text{DET}(E_n)\text{DET}(E_{n-1}) \ldots \text{DET}(E_2) \, \text{DET}(E_1) \, \text{DET}(B) = \text{DET}(AB) \qquad (4.3)$$

But using the same result repeatedly on Equation (4.2) above we see that

$$\text{DET}(E_n) \, \text{DET}(E_{n-1}) \ldots \text{DET}(E_2) \, \text{DET}(E_1) = \text{DET}(A)$$

Hence (4.3) gives us the required result, namely,

$$\text{DET}(A) \, \text{DET}(B) = \text{DET}(AB)$$

Before we leave the topic of elementary matrices we observe another fact about them.

Since the determinant of an elementary matrix is $+1$, -1 or $k \neq 0$, it follows that **all** elementary matrices have inverses. Further, the inverse of an elementary matrix is another elementary matrix of the same type.

To write down a matrix A as a product of elementary matrices, we needed A to have an inverse. We can generalize this statement to the following fact.

Any $n \times n$ matrix A can be written as a product of elementary matrices together with a matrix U which is upper triangular.

Note that we could replace U with a matrix L which is lower triangular. The way this comes about is that we start with the matrix A and row-reduce it until we have a lower or upper triangular matrix. This we can always achieve. The elementary row operations we perform to arrive at this state can be represented by elementary matrices, $E_n, E_{n-1}, \ldots, E_2, E_1$. Hence we can write for **any** $n \times n$ matrix A

$$E_n E_{n-1} \ldots E_2 E_1 A = U$$

and hence,

$$A = E_1^{-1} E_2^{-1} \cdots E_{n-1}^{-1} E_n^{-1} U$$

Example 4D

Let

$$A = \begin{bmatrix} 1 & 2 & -1 \\ 0 & 3 & 1 \\ 2 & 1 & -1 \end{bmatrix}$$

Write A as a product of elementary matrices and an upper triangular matrix.

Solution

Recall that an upper triangular matrix has all its non-zero elements either on the leading diagonal or above it. In our case we need to zero the elements in positions a_{31} and a_{32}.

The elementary row operation, $R_3 \rightarrow R_3 - 2R_1$ results in the matrix

$$\begin{bmatrix} 1 & 2 & -1 \\ 0 & 3 & 1 \\ 0 & -3 & 1 \end{bmatrix}$$

The elementary matrix giving this matrix by multiplication $E_1 A$ is

$$E_1 = \begin{bmatrix} 1 & 0 & 0 \\ 0 & 1 & 0 \\ -2 & 0 & 1 \end{bmatrix}$$

Next we perform the elementary operation $R_3 \rightarrow R_3 + R_2$ on the matrix to obtain the matrix

$$\begin{bmatrix} 1 & 2 & -1 \\ 0 & 3 & 1 \\ 0 & 0 & 2 \end{bmatrix}$$

This is an upper triangular matrix and is the matrix U we are looking for.
The elementary matrix which gives us the same result would be

$$E_2 = \begin{bmatrix} 1 & 0 & 0 \\ 0 & 1 & 0 \\ 0 & 1 & 1 \end{bmatrix}$$

Hence in this exercise

$$E_2 E_1 A = U$$

Therefore, by multiplying both sides on the left by E_2^{-1}, which we know must exist, we obtain

$$E_1 A = E_2^{-1} U$$

Finally we multiply both sides on the left by E_1^{-1} to obtain

$$A = E_1^{-1} E_2^{-1} U$$

Hence we have achieved the desired result; A is indeed a product of elementary matrices and an upper triangular matrix.

Exercise 4G

Write the following matrix as a product of elementary matrices and an upper triangular matrix:

$$\begin{bmatrix} 1 & -1 & 4 \\ 2 & 0 & 1 \\ 3 & 1 & 2 \end{bmatrix}$$

DERIVE ACTIVITY 4F

Check the result of Example 4D and Exercise 4G using DERIVE.

Use DERIVE to write down the matrices E_2^{-1} and E_1^{-1} explicitly.

Start with the same matrix A and reduce it to a product of elementary matrices together with a lower triangular matrix. You will have to consider your strategy for doing this carefully, otherwise you will have zero entries turning back into non-zero entries in your row reductions.

Exercise 4H

Consider the 3×4 matrix

$$A = \begin{bmatrix} 1 & -2 & 3 & -1 \\ 2 & -1 & 2 & 2 \\ 3 & 1 & 2 & 3 \end{bmatrix}$$

This matrix can be considered as a 3×3 matrix augmented with a 3×1 column matrix.

Perform elementary row operations on A to bring it into row-reduced echelon form.

Hint. Here you should be trying to obtain A in the form $[I \mid B]$, where I is the 3×3 identity matrix and B is a 3×1 column matrix.

In your solution you should have obtained the following matrix after eight operations:

$$\begin{bmatrix} 1 & 0 & 0 & 15/7 \\ 0 & 1 & 0 & -4/7 \\ 0 & 0 & 1 & -10/7 \end{bmatrix}$$

Note that in general we may not always be able to obtain the identity matrix in the left-hand portion of the augmented matrix.

It is at such times in applications that life becomes interesting!

We are now in a position to study systems of equations such as the following set of *linear equations in three variables*:

$$x - 2y + 3z = -1$$

$$2x - y + 2z = 2$$

$$3x + y + 2z = 3$$

Notice these have some similarity to the matrix in the last exercise. If we just consider the coefficients and the constants on the right-hand side of the equations, we obtain the 3 × 4 matrix A.

Before we bring the different ideas together we introduce one final concept in this chapter.

4.8   THE RANK OF A MATRIX

Recall the concept of vectors being linearly independent, which we introduced in Chapter 3. The rows of any matrix can be interpreted as vectors (as can the columns). Hence we can talk of linearly independent rows in a matrix.

Definitions

The number of linearly independent rows of a matrix is called the *row rank of the matrix*. Similarly, the number of independent columns of a matrix is called *the column rank of a matrix*.

There is a theorem in linear algebra that states that the row rank of a matrix equals the column rank of a matrix. A proof of this fact can be found in most traditional text books. Because of this theorem, we refer to the rank of a matrix as the number of linearly independent rows or columns of a matrix. In practice, to obtain the rank of a matrix is straightforward if we use the idea of row reduction introduced earlier. To illustrate the ideas consider Exercise 4H and the row-reduced echelon form.

We had a 3 × 4 matrix which was reduced to the matrix given below:

$$\begin{bmatrix} 1 & 0 & 0 & 15/7 \\ 0 & 1 & 0 & -4/7 \\ 0 & 0 & 1 & -10/7 \end{bmatrix}$$

If we now consider the three rows as vectors with four entries in each, we ask the question:

How many of these three vectors are linearly independent?

Whatever the answer to this question is, be it 0, 1, 2 or 3, that will be the rank of the matrix and therefore the rank of the original matrix. This follows from the fact that the rank of a matrix A is equal to the rank of any other matrix obtained from A by elementary row operations.

We will call row 1, v_1, row 2 v_2, and row 3 v_3 in our discussion below.

By inspection the following pairs of rows are independent: v_1 and v_2; v_1 and v_3; v_2 and v_3. That is, we *cannot* obtain a relation $v_i = \alpha v_j$ for any i and j in $\{1, 2, 3\}$.

It follows, then, that v_1, v_2 and v_3 are linearly independent and therefore the rank of the original matrix is 3.

In practice we can always reduce a matrix to echelon form and then simply count the non-zero rows. This will give us the rank of the matrix.

Recall that zero rows, if they occur, will be in the bottom rows of the reduced matrix. Furthermore, if zero rows appear in the reduction process, it will mean that that particular row has been eliminated by a linear combination of rows higher up in the matrix.

Let us, for argument's sake, say that v_i is a zero row after reduction. Then we must have a relationship of the form

$$v_i = \alpha_1 v_1 + \alpha_2 v_2 + \alpha_3 v_3 + ... + \alpha_{i-1} v_{i-1}$$

Remember, we are counting rows from the top of the matrix.

In the example above no zero rows can be obtained so, as stated, the rank is 3.

If we reduce the same matrix using columns we would obtain one column full of zeros showing its column rank to be 3. This would verify that

'column rank = row rank'

The following example illustrates this result.

Example 4E

Consider again the 3×4 matrix from Exercise 4H:

$$A = \begin{bmatrix} 1 & -2 & 3 & -1 \\ 2 & -1 & 2 & 2 \\ 3 & 1 & 2 & 3 \end{bmatrix}$$

Reduce A by column operations.

Solution

Label the columns c_1, c_2, c_3 and c_4. Using c_1, column 1, we can make a_{12}, a_{13} and a_{14} all equal to zero by the following operations:

$$c_2 \rightarrow c_2 + 2c_1 \qquad c_3 \rightarrow c_3 - 3c_1 \qquad c_4 \rightarrow c_4 + c_1$$

We now have the matrix

$$\begin{bmatrix} 1 & 0 & 0 & 0 \\ 2 & 3 & -4 & 4 \\ 3 & 7 & -7 & 6 \end{bmatrix}$$

The next operation is to divide column 2, c_2, by 3 to obtain $a_{22} = 1$:

$$\begin{bmatrix} 1 & 0 & 0 & 0 \\ 2 & 1 & -4 & 4 \\ 3 & 7/3 & -7 & 6 \end{bmatrix}$$

We now use this new entry to make a_{21}, a_{23} and a_{24} all equal to zero, by the following operations:

$$c_1 \rightarrow c_1 - 2c_2 \qquad c_3 \rightarrow c_3 + 4c_2 \qquad c_4 \rightarrow c_4 - 4c_2$$

The resulting matrix is given below:

$$\begin{bmatrix} 1 & 0 & 0 & 0 \\ 0 & 1 & 0 & 0 \\ -5/3 & 7/3 & 7/3 & -10/3 \end{bmatrix}$$

Next we divide column 3 by 7/3 to make $a_{33} = 1$:

$$\begin{bmatrix} 1 & 0 & 0 & 0 \\ 0 & 1 & 0 & 0 \\ -5/3 & 7/3 & 1 & -10/3 \end{bmatrix}$$

Finally, we use this new entry to make a_{31}, a_{32} and a_{34} all equal to zero. The final matrix would then be

$$\begin{bmatrix} 1 & 0 & 0 & 0 \\ 0 & 1 & 0 & 0 \\ 0 & 0 & 1 & 0 \end{bmatrix}$$

Hence we do have a column of zeros and the first three columns are linearly independent. Thus the column rank is 3, the same as the row rank.

Exercise 4I

Two matrices are given below.

Using the ideas introduced above, show that the rank of A is 2 and the rank of B is 3:

$$A = \begin{bmatrix} 2 & -1 & -2 \\ 1 & 1 & 1 \\ 4 & 10 & 12 \end{bmatrix} \qquad B = \begin{bmatrix} 2 & 1 & 2 \\ -1 & 1 & 2 \\ -2 & 2 & 1 \\ 1 & 2 & 3 \end{bmatrix}$$

DERIVE ACTIVITY 4G

(A)   DERIVE did **not** have a function Rank(A) or Rk(A) in its earlier version. This was no disadvantage. It did, however, have an instruction which row-reduced a matrix to what we have called reduced echelon form earlier in this chapter, and, as we have seen in the last few exercises, we can read off the rank of a matrix by counting the number of non-zero rows.

Load DERIVE and from Exercise 4I input matrix A into line 1 and matrix B into line 2:

Author ROW_REDUCE(#1)

Now **Simplify** this expression:

Author ROW_REDUCE(#2) and **Simplify**

Your screen should look like that shown in Figure 4.1.

#3: ROW_REDUCE $\begin{bmatrix} 2 & -1 & -2 \\ 1 & 1 & 1 \\ 4 & 10 & 12 \end{bmatrix}$

#4: $\begin{bmatrix} 1 & 0 & -\dfrac{1}{3} \\ 0 & 1 & \dfrac{4}{3} \\ 0 & 0 & 0 \end{bmatrix}$

#5: ROW_REDUCE $\begin{bmatrix} 2 & 1 & 2 \\ -1 & 1 & 2 \\ -2 & 2 & 1 \\ 1 & 2 & 3 \end{bmatrix}$

#6: $\begin{bmatrix} 1 & 0 & 0 \\ 0 & 1 & 0 \\ 0 & 0 & 1 \\ 0 & 0 & 0 \end{bmatrix}$

COMMAND: **Author** Build Calculus Declare Expand Factor Help Jump soLve Manage
         Options Plot Quit Remove Simplify Transfer Unremove moVe Window approX
Enter option
Simp(#5)   Free:74%   Derive Algebra

Figure 4.1   Using ROW_REDUCE to find the Rank

(B)   Further, in this activity let us 'keep track' of the matrices we obtain as we step-
      by-step row-reduce our original matrix.
            DERIVE helps us to do this with functions in VECTOR.MTH.
      Transfer, Load, Derive, VECTOR.MTH, the functions are loaded line-by-
      line so you can actually see them on your screen. While this can be useful to
      remind you how to use a certain function, it does have the disadvantage of
      taking up space in the memory. If you wish to save space then use **Transfer,
      Load, Utility,** as we have already done earlier in the book.
      Now let us return to our matrix *A*, at the beginning of this activity. We are
      going to row-reduce it step-by-step to obtain the matrix on line 4 of Figure 4.1.
      There is no unique way of doing this, although of course the final result is
      unique!

Author matrix A in line 1.

The new function we introduce here is

SCALE_ELEMENT (#1,1,1/2)

This takes the matrix in line 1, finds the first row and multiplies each of its entries by $\frac{1}{2}$. The result is that element a_{11} becomes 1, and we can then use it to zero the elements in positions, a_{21} and a_{31}.

Author the above expression; it will appear on line 2 and then **Simplify** will give you the reduced matrix in line 3. Your screen should look like that shown in Figure 4.2.

#1: $\begin{bmatrix} 2 & -1 & -2 \\ 1 & 1 & 1 \\ 4 & 10 & 12 \end{bmatrix}$

#2: $\text{SCALE\_ELEMENT}\left[\begin{bmatrix} 2 & -1 & -2 \\ 1 & 1 & 1 \\ 4 & 10 & 12 \end{bmatrix}, 1, \dfrac{1}{2}\right]$

#3: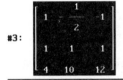

COMMAND: **Author** Build Calculus Declare Expand Factor Help Jump soLve Manage
          Options Plot Quit Remove Simplify Transfer Unremove moVe Window approX
Compute time: 0.1 seconds
Simp(#2)   Free:74%   Derive Algebra

Figure 4.2   SCALE_ELEMENT multiplies row 1 by ½

To complete this task will take about another 16 lines in DERIVE.

If you follow the instructions given below, you should see the reduced matrix on line 19 as it appears in Figure 4.1. Before we list the instructions for you we introduce the final function that we need to reduce the matrix A further.

When you type

SUBTRACT_ELEMENTS(#3,2,1)

DERIVE will take the matrix in line 3 and subtract row 1 from row 2. **Author** this statement and then **Simplify** (lines 4 and 5).

Now **Author**

SCALE_ELEMENT(#5,3,1/4)

and **Simplify** (lines 6 and 7). This has the effect of making the element in position $a_{31} = 1$.

Now **Author**

SUBTRACT_ELEMENTS(#7,3,1)

in line 8 and **Simplify** to give a matrix in line 9 with zeros in positions a_{31} and a_{21}.

Now we will scale the elements in row 3 by a factor of $\frac{1}{2}$ to reveal two identical rows.

Author

SCALE_ELEMENT(#9,3,1/2)

in line 10 and **Simplify** to give a matrix in line 11 with row 2 and row 3 identical.

Now **Author**

SUBTRACT_ELEMENTS(#11,3,2)

on line 12 and **Simplify** to get on line 13 a matrix with row three full of zeros.
On line 14, **Author**

SCALE_ELEMENT(#13,2, −1/3)

to make elements a_{12} and a_{22} equal to $-\frac{1}{2}$. The simplified matrix should appear on line 15.
On line 16 **Author**

SUBTRACT_ELEMENTS(#15,1,2)

to make $a_{12} = 0$ after simplifying on line 17.
Finally, on line 18, **Author**

SCALE_ELEMENT(#17,2,−2)

and **Simplify** on line 19 to give the reduced matrix we found at line 4 in Figure 4.1.

It has taken quite a while to use the SCALE_ELEMENT and SUBTRACT_ELEMENTS functions from VECTOR.MTH when you consider that ROW_REDUCE gives the same result in one line. However, if you have completed this activity you should feel confident in using the two functions introduced here to manipulate matrices to whatever form is asked for. In many cases you do not need or want a row-reduced form.

(C)   Use DERIVE to find the rank of each of the following matrices:

$$\begin{bmatrix} 1 & 3 & -1 \\ 2 & 1 & 0 \\ 1 & 4 & -1 \end{bmatrix} \qquad \begin{bmatrix} 2 & -1 & 3 & 1 \\ 1 & 4 & 2 & -1 \\ -1 & 0 & 1 & -2 \end{bmatrix} \qquad \begin{bmatrix} 1 & 4 \\ 2 & -1 \\ 3 & 0 \\ 1 & 5 \end{bmatrix}$$

The rank of a matrix plays an important role in the solution of systems of linear equations, as we will see. In the next chapter we will bring together different ideas from preceding chapters to show that we have numerous ways of attacking a system of linear equations to obtain a solution, whether it is unique or not.

Consolidation Exercise

1.   (a)   For what value(s) of t will the matrix given below **not** have an inverse?

$$\begin{bmatrix} 1 & t & 0 \\ 0 & 1 & -1 \\ t & 0 & 1 \end{bmatrix}$$

      (b)   For all other values of t, other than those found in (a), work out the inverse of A.
2.   Verify that if A is an idempotent matrix, then so is the matrix $(I-A)$, where I is the appropriate identity matrix.
            Now consider the matrix $(I - 2A)$.
            What is its inverse?
            Is $(I - 2A)$ idempotent?
            What, if anything, can be said about the matrix $(I - 3A)$?
3.   Given that two matrices A and B are such that

$$A^2 = AB \quad \text{and} \quad A \neq 0, \quad \text{the zero matrix}$$

Find the error in the following argument.

To prove $A = B$.

Since $A^2 = AB$ then $A^2 - AB = 0$.
    Factorizing this gives $A(A - B) = 0$.
    Since $A \neq 0$, $A - B = 0$.
    Hence $A = B$.

4.   Show that the vectors [1 0 1], [–1 0 1] and [0 1 0] are mutually orthogonal in $\Re^3$.

    Write down the 3×3 matrix which has the above three vectors for its columns and call this matrix A.

    Work out the products, AA' and $A'A$, where A' means the transpose of A. Hence show that these products can be written in the form

$$\text{diag}(a \quad b \quad c)$$

Write down the values a, b and c for each product.
    Does A commute with its transpose?
    Show that A has an inverse and write it down.
    If we began this problem again with any three mutually orthogonal vectors from $\Re^3$ would we get the same results?

5.   Find the inverse of the 4×4 matrix given below:

$$\begin{bmatrix} 1 & 1 & 2 & 3 \\ 0 & 1 & 1 & 2 \\ 0 & 0 & 1 & 1 \\ 0 & 0 & 0 & 1 \end{bmatrix}$$

(i)   by forming the adjugate or adjoint matrix and computing the determinant, by hand,
(ii)   by using (5) from Exercise 1D of Chapter 1,
(iii)   by using simple row operations on the appropriate augmented matrix,
(iv)   by using DERIVE to check your answer!

5

Systems of Linear Equations

Systems of equations arise in many disciplines, for example, engineering, economics and modelling. The equations arise from everyday situations and they need solutions if, for example, manufacturing industry and economic forecasting are to progress. In this chapter we will begin to solve systems of linear equations by hand, using DERIVE to assist us when the going becomes difficult. We will use many, if not all, the techniques and ideas introduced in the earlier chapters.

5.1   RECAP FROM PRECEDING CHAPTERS

At the beginning of Chapter 2, we had a system of two equations in two unknowns and, a little later, three equations in three unknowns. Such systems are said to be *square*. In general, if we have n linear equations in n unknowns we would refer to this as the $n \times n$ *square system*. We can write such systems in matrix form as follows:

$$A\mathbf{x} = \mathbf{k}$$

Here the matrix A is the $n \times n$ matrix of coefficients. The matrix $\mathbf{x}$ is $n \times 1$, a column vector with entries $x_1, x_2, ..., x_n$, the n unknowns. The matrix $\mathbf{k}$ is another $n \times 1$ column vector which usually has constant entries.

An example of a 2×2 system with unknowns x_1 and x_2 is given below:

$$2x_1 + 3x_2 = 6$$
$$-x_1 - 2x_2 = 5$$

This can be written in matrix form:

$$\begin{bmatrix} 2 & 3 \\ -1 & -2 \end{bmatrix} \begin{bmatrix} x_1 \\ x_2 \end{bmatrix} = \begin{bmatrix} 6 \\ 5 \end{bmatrix}$$

Compare this to

$$A\mathbf{x} = \mathbf{k}$$

Our task is to 'solve' the system $A\mathbf{x} = \mathbf{k}$, as we stated in the introduction above, that is, find $\mathbf{x}$.

In Chapter 2 we introduced Cramer's rule. This was a way of obtaining the unique solution to the system (if it had one). The unique solution relied on the fact that $\text{DET}(A) \neq 0$. We call this method of solving the system, *solution by determinants*. We also saw in Chapter 4 that the inverse of a square matrix A exists if $\text{DET}(A) \neq 0$. Thus, provided that A^{-1} exists, that is, $\text{DET}(A) \neq 0$, the equations can be solved formally in the following way.

Starting with our system in matrix form we have

$$A\mathbf{x} = \mathbf{k}$$

Since A^{-1} exists we can multiply both sides of this equation on the left to obtain

$$A^{-1}A\mathbf{x} = A^{-1}\mathbf{k}$$

Using the result $AA^{-1} = I$ (the identity matrix), this equation reduces to

$$\mathbf{x} = A^{-1}\mathbf{k}$$

Hence if there is a solution, it can be read off by comparing elements of the left-hand side, that is, $x_1, x_2, ..., x_n$, with elements of the right-hand side.

Example 5A

Solve the following system of equations:

$$\begin{aligned}
x_1 - 2x_2 + 3x_3 &= -1 \\
2x_1 - x_2 + 2x_3 &= 2 \\
3x_1 + x_2 + 2x_3 &= 3
\end{aligned}$$

Solution

Here

$$\mathbf{x} = \begin{bmatrix} x_1 \\ x_2 \\ x_3 \end{bmatrix} \quad \text{and} \quad \mathbf{k} = \begin{bmatrix} -1 \\ 2 \\ 3 \end{bmatrix} \quad \text{and} \quad A = \begin{bmatrix} 1 & -2 & 3 \\ 2 & -1 & 2 \\ 3 & 1 & 2 \end{bmatrix}$$

To see if A^{-1} exists we calculate DET(A):

$$\text{DET}(A) = 1\begin{vmatrix} -1 & 2 \\ 1 & 2 \end{vmatrix} + 2\begin{vmatrix} 2 & 2 \\ 3 & 2 \end{vmatrix} + 3\begin{vmatrix} 2 & -1 \\ 3 & 1 \end{vmatrix} = 7$$

Using the adjugate matrix to find A^{-1} we have

$$A^{-1} = \frac{1}{7}\begin{bmatrix} -4 & 2 & 5 \\ 7 & -7 & -7 \\ -1 & 4 & 3 \end{bmatrix}^T = \frac{1}{7}\begin{bmatrix} -4 & 7 & -1 \\ 2 & -7 & 4 \\ 5 & -7 & 3 \end{bmatrix}$$

Hence, calculating $A^{-1}k$ gives

$$\mathbf{x} = \frac{1}{7}\begin{bmatrix} -4 & 7 & -1 \\ 2 & -7 & 4 \\ 5 & -7 & 3 \end{bmatrix}\begin{bmatrix} -1 \\ 2 \\ 3 \end{bmatrix} = \begin{bmatrix} 15/7 \\ -4/7 \\ -10/7 \end{bmatrix}$$

Exercise 5A

Solve the linear equations in Example 5A using Cramer's rule.

Using DERIVE we can quickly obtain the value of the determinant and also the inverse.

DERIVE ACTIVITY 5A

(A)   Load DERIVE and input matrix A, from Example 5A.

Show that the determinant is non-zero and thus the inverse exists and is given by

$$A^{-1} = \frac{1}{7} \begin{bmatrix} -4 & 7 & -1 \\ 2 & -7 & 4 \\ 5 & -7 & 3 \end{bmatrix}$$

Finally, in this exercise use DERIVE to find the solution. Your screen after this exercise should look similar to that shown in Figure 5.1.

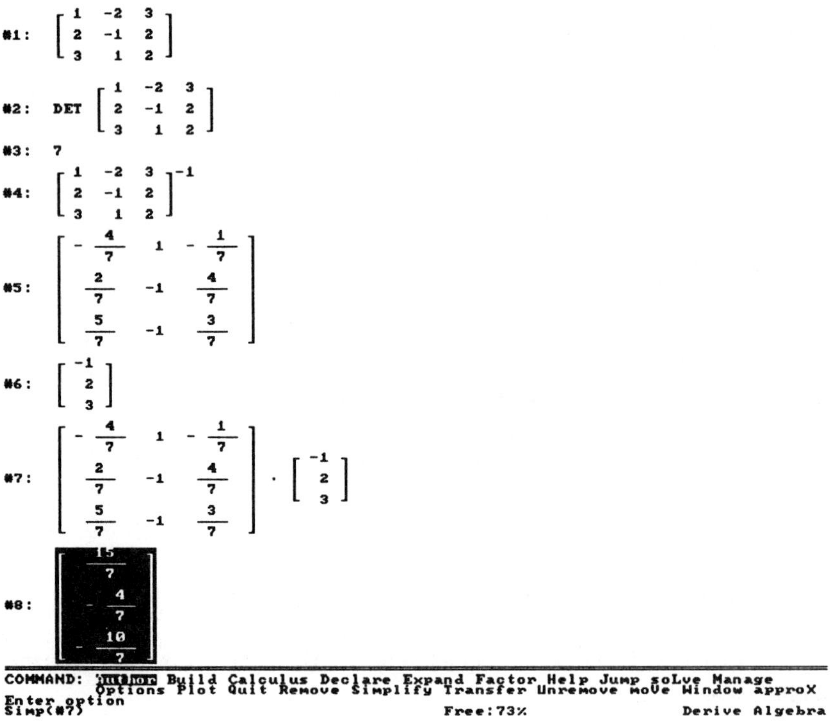

#1: $\begin{bmatrix} 1 & -2 & 3 \\ 2 & -1 & 2 \\ 3 & 1 & 2 \end{bmatrix}$

#2: DET $\begin{bmatrix} 1 & -2 & 3 \\ 2 & -1 & 2 \\ 3 & 1 & 2 \end{bmatrix}$

#3: 7

#4: $\begin{bmatrix} 1 & -2 & 3 \\ 2 & -1 & 2 \\ 3 & 1 & 2 \end{bmatrix}^{-1}$

#5: $\begin{bmatrix} -\dfrac{4}{7} & 1 & -\dfrac{1}{7} \\ \dfrac{2}{7} & -1 & \dfrac{4}{7} \\ \dfrac{5}{7} & -1 & \dfrac{3}{7} \end{bmatrix}$

#6: $\begin{bmatrix} -1 \\ 2 \\ 3 \end{bmatrix}$

#7: $\begin{bmatrix} -\dfrac{4}{7} & 1 & -\dfrac{1}{7} \\ \dfrac{2}{7} & -1 & \dfrac{4}{7} \\ \dfrac{5}{7} & -1 & \dfrac{3}{7} \end{bmatrix} \cdot \begin{bmatrix} -1 \\ 2 \\ 3 \end{bmatrix}$

#8: $\begin{bmatrix} \dfrac{15}{7} \\ -\dfrac{4}{7} \\ -\dfrac{10}{7} \end{bmatrix}$

COMMAND: **Author** Build Calculus Declare Expand Factor Help Jump soLve Manage
         Options Plot Quit Remove Simplify Transfer Unremove moVe Window approX
Enter option
Simp(#7)   Free:73%   Derive Algebra

Figure 5.1 Solving linear equations using DERIVE

Hence,

$$x_1 = 15/7, \quad x_2 = -4/7 \quad \text{and} \quad x_3 = -10/7$$

is the unique solution to this system. We will call this method the *solution by inverses*.

(B)   In Chapter 4, Exercise 4H, you were asked to row-reduce a 3×4 matrix. If you examine this matrix you will observe that it is made up of the matrix of coefficients and **k** from this activity. Furthermore, if you examine the reduced matrix you will find the identity matrix augmented with **x**. This is not a coincidence, as we will now show.

Using the matrices A, **k** and **x** given at the start of this activity write down the augmented matrix

$$(A|I|\mathbf{k})$$

Now use DERIVE to row-reduce this matrix. You should obtain the augmented matrix

$$(I|A^{-1}|\mathbf{x})$$

What has happened is that the elementary matrices have operated not just on A and I but also on **k** and have produced I and A^{-1}, as we saw in Chapter 4, together with **x**, the unique solution to our system.

(C)   Repeat activity (B) for the matrices with appropriate **x** and **k**

$$\begin{bmatrix} -3 & 7 & 2 \\ 4 & 0 & -5 \\ 1 & -1 & 6 \end{bmatrix} \qquad \begin{bmatrix} -4 & 6 & 1 & 2 \\ 2 & -1 & 1 & 3 \\ 7 & 5 & 3 & 0 \\ -1 & 2 & 4 & 1 \end{bmatrix}$$

5.2 NON-UNIQUE SOLUTIONS

Thus far, we have obtained unique solutions because $\text{DET}(A) \neq 0$. The next stage is to consider what happens if $\text{DET}(A) = 0$. The first observation is that no unique solution can exist to our system. Can we obtain any solutions? The answer to this question is that sometimes there are no solutions and sometimes there are more solutions than we can cope with. We then say that we have an infinity of solutions.

To begin to study this situation let us go to the simplest case possible, that of one equation in one unknown:

$$ax_1 = b$$

Here we have a 1×1 situation, the matrix $A = [a]$, $\mathbf{x} = [x_1]$ and $\mathbf{k} = [b]$.

First, if $a \neq 0$, then there is a **unique** solution for x_1 namely b/a. However, if $a = 0$, then we have two possible cases to consider. Since b is a constant it is either zero or it is not zero. This is a simple truth about any number.

Case 1

Let $a = 0$ and $b = 0$. We now have

$$0 \cdot x_1 = 0$$

Hence, we can have any value for x_1. In this case we have an *infinity of solutions*.

Case 2

Let $a = 0$ and $b \neq 0$. We now have

$$0 \cdot x_1 = b \neq 0$$

This comes down to saying

$$0 \neq 0$$

In this case we say that the equation is *inconsistent*. In this case *no solution exists*.

Solutions and Ranks

We will give, without proof, the theorems which enable us to know whether we have a unique solution, an infinity of solutions or no solutions by using the idea of the rank of a matrix introduced at the end of the last chapter. Again let us consider the $n \times n$ system $A\mathbf{x} = \mathbf{k}$.

Criterion for a Unique Solution

> The system of equations $A\mathbf{x} = \mathbf{k}$ has a unique solution if and only if the rank of the matrix A, rank(A), is equal to n.

This is reasonable since all the rows of A are independent if Rank(A) = n and thus DET(A) $\neq 0$ and hence A^{-1} exists and a unique solution follows.

Criterion for Non-unique Solutions

> If Rank(A) = Rank(A|**k**) then the system has an infinity of solutions but if Rank(A) ≠ Rank(A|**k**) then the system is inconsistent and no solutions exist.
>   Note that Rank(A) < n in these cases.

We give below five cases where we have a 2×2 system and various outcomes. The equations are kept simple so that you can do all the calculations by hand and thus see what is happening.

(i)   Consider the following two equations:

$$x_1 + x_2 = 2$$
$$x_1 - x_2 = 0$$

As before, the system can be written in matrix form as

$$\begin{bmatrix} 1 & 1 \\ 1 & -1 \end{bmatrix} \cdot \begin{bmatrix} x_1 \\ x_2 \end{bmatrix} = \begin{bmatrix} 2 \\ 0 \end{bmatrix}$$

This in turn is equivalent to the general equation

$$A \cdot \mathbf{x} = \mathbf{k}$$

Rank(A) = 2 since an initial row reduction of A gives the matrix

$$\begin{bmatrix} 1 & 1 \\ 0 & -2 \end{bmatrix}$$

which then reduces further to

$$\begin{bmatrix} 1 & 0 \\ 0 & 1 \end{bmatrix}$$

This matrix cannot be reduced further as it is now the identity matrix and the two rows regarded as vectors are independent. In fact the two rows are the two-dimensional vectors **i** and **j** introduced earlier.
     Thus using our criterion when Rank(A) = size of matrix, n, we have a unique solution

$$x_1 = 1 \quad \text{and} \quad x_2 = 1$$

This solution could be obtained by considering the original equations simultaneously or by row-reducing the matrix $[A|\mathbf{k}]$ to $[I|\mathbf{x}]$, as discussed in Section 5.1 above.

(ii)   Now consider the system

$$x_1 + x_2 = 2$$
$$x_1 + x_2 = 1$$

Following the same reasoning as in the first example we row-reduce the matrix

$$\begin{bmatrix} 1 & 1 \\ 1 & 1 \end{bmatrix}$$

and obtain

$$\begin{bmatrix} 1 & 1 \\ 0 & 0 \end{bmatrix}$$

This time we obtain a row of zeros and thus the rank of the original matrix is 1. However, $n = 2$ again as we have a 2×2 system.

Now consider the rank of $[A|\mathbf{k}]$, which is displayed below:

$$\begin{bmatrix} 1 & 1 & 2 \\ 1 & 1 & 1 \end{bmatrix}$$

This reduces to

$$\begin{bmatrix} 1 & 1 & 2 \\ 0 & 0 & -1 \end{bmatrix}$$

and finally to

$$\begin{bmatrix} 1 & 1 & 0 \\ 0 & 0 & 1 \end{bmatrix}$$

The rank of $[A|\mathbf{k}]$ is 2 by inspection.

Using the criterion that

$$\text{Rank}(A) \neq \text{Rank}(A|\mathbf{k})$$

we have inconsistency and thus this system has no solutions.

We remark that it was reasonably clear at the outset that we had two equations with the same left-hand sides but with differing right-hand sides. Therefore, we couldn't expect any consistent solutions.

(iii)   Now consider the system

$$x_1 + x_2 = 2$$
$$2x_1 + 2x_2 = 4$$

Here

$$A = \begin{bmatrix} 1 & 1 \\ 2 & 2 \end{bmatrix}$$

and

$$[A|\mathbf{k}] = \begin{bmatrix} 1 & 1 & 2 \\ 2 & 2 & 4 \end{bmatrix}$$

Again Rank(A) = 1 since a row of zeros can be obtained by the row operation

$$R_2 \rightarrow R_2 - 2R_1$$

Furthermore, Rank($A|\mathbf{k}$) = 1, since performing the same row operation on the augmented matrix produces a row of zeros. Hence,

$$\text{Rank}(A) = \text{Rank}(A|\mathbf{k}) = 1 < 2 = n$$

and therefore, by the criteria above, we have an infinity of solutions. These solutions can in fact be given as follows:

If $x_1 = a$ then from either equation $x_2 = 2 - a$

As 'a' can take any real value it follows that an infinity of solutions for x_1 and x_2 exists.

(iv)   Now consider the system given below:

$$x_1 + x_2 = 2$$
$$x_1 - \alpha x_2 = \alpha$$

Here

$$A = \begin{bmatrix} 1 & 1 \\ 1 & -\alpha \end{bmatrix}$$

and

$$(A|\mathbf{k}) = \begin{bmatrix} 1 & 1 & 2 \\ 1 & -\alpha & \alpha \end{bmatrix}$$

The matrix A reduces to

$$\begin{bmatrix} 1 & 1 \\ 0 & -\alpha - 1 \end{bmatrix}$$

Now we consider possible cases depending on the value of α.
    If $\alpha \neq -1$, Rank(A) = 2 and a unique solution exists.
    If $\alpha = -1$ then Rank(A) = 1.
    Furthermore, the matrix $(A|\mathbf{k})$ is

$$(A|\mathbf{k}) = \begin{bmatrix} 1 & 1 & 2 \\ 1 & 1 & -1 \end{bmatrix}$$

which reduces by the row operation $R_2 \rightarrow R_2 - R_1$ to the matrix

$$\begin{bmatrix} 1 & 1 & 2 \\ 0 & 0 & -3 \end{bmatrix}$$

and hence Rank($A|\mathbf{k}$) = 2.
    The equations are inconsistent by the criteria and thus no solutions exist.

(v)   Finally consider the system of equations given below:

$$x_1 + x_2 = -1$$
$$x_1 - \alpha x_2 = \alpha$$

Using some of the results from (iv), Rank(A) = 2 if $\alpha \neq -1$. Thus a unique solution exists, in this case. However, if $\alpha = -1$, then Rank(A) and Rank($A|\mathbf{k}$) are both equal to 1.

Therefore, an infinity of solutions exists. These can be obtained by choosing $x_1 = a$ and hence from either equation we obtain $x_2 = -1 - a$.

5.3   HOMOGENEOUS EQUATIONS

So far, we have been considering the $n \times n$ system of equations given in general form by the matrix equation $A\mathbf{x} = \mathbf{k}$. Most of our examples have been for $n = 2$, although the same ideas apply for any n.

Definition

The system of equations $A\mathbf{x} = \mathbf{0}$ is said to be *homogeneous*. This is a special case of our general equation where the constant vector $\mathbf{k}$ is simply the zero vector of size $n \times 1$. By inspection this system will always have at least one solution.

That solution is $\mathbf{x} = \mathbf{0}$, which we call *the trivial solution*. In simple terms it means that $x_1 = x_2 = ... = x_n = 0$.

Alternatively we can see this solution from the following reasoning. If Rank(A) = n, then A^{-1} exists and from the equation $A\mathbf{x} = \mathbf{0}$ we obtain the equation $\mathbf{x} = A^{-1} \cdot \mathbf{0} = \mathbf{0}$ and hence $x_i = 0, \forall i$.

If Rank(A) < n we observe that Rank($A|0$) = Rank(A) so there must be an *infinity of solutions*. To find these solutions we use determinants and their properties.

Example 5B

Find all the solutions of the equations

$$x_1 + 2x_2 = 0$$
$$3x_1 + \alpha x_2 = 0$$

Solution

First consider the matrix of coefficients

$$A = \begin{bmatrix} 1 & 2 \\ 3 & \alpha \end{bmatrix}$$

Furthermore,

$$\mathrm{DET}(A) = \alpha - 6$$

Now if $\alpha = 6$ then $\mathrm{DET}(A) = 0$. In this case there is an infinity of solutions, since the second row of A is simply three times the first row and thus, in reduction, a row of zeros will be obtained. These solutions can be given by $x_1 = a$ and $x_2 = -a/2$.

However if $\alpha \neq 6$ then $\mathrm{DET}(A) \neq 0$ and thus A^{-1} exists and the only solution is the trivial one.

Exercise 5B

Investigate the possible solutions of the following system of homogeneous equations:

$$(1-\lambda)x_1 + 2x_2 = 0$$

$$2x_1 + (1-\lambda)x_2 = 0$$

5.4   GAUSSIAN ELIMINATION

We now discuss an algorithmic method of solving systems of equations for the case where solutions exist. The method is essentially one of systematically eliminating unknown variables until an equivalent set of equations in triangular form is obtained. The triangular set of equations is then solved by back substitution. We will introduce this idea with an example.

Consider again the system of equations introduced in the preceding chapter:

$$
\begin{aligned}
x - 2y + 3z &= 1 \\
2x - y + 2z &= -2 \\
3x + y + 2z &= -3
\end{aligned}
$$

If we call the matrix of coefficients A, then the augmented matrix $(A|k)$ will be as given below:

$$
\left[
\begin{array}{ccc|c}
1 & -2 & 3 & 1 \\
2 & -1 & 2 & -2 \\
3 & 1 & 2 & -3
\end{array}
\right]
$$

We now proceed to reduce this matrix by hand, step-by-step, to show the Gauss elimination procedure in action. (Using DERIVE we could obtain the result very quickly. We will do this later in this chapter.)

We begin the row reduction by taking the x term in the first equation, and using it we eliminate the x terms from the second and third equations. Remember that we start with a system of equations and we obtain equivalent systems of equations by performing elementary row operations on the equations. To do this we only need to keep track of the coefficients and the constants as displayed in the augmented matrix above. At any stage we could write the equivalent equations down.

We perform the following two row operations:

$$R_2 \rightarrow R_2 - 2R_1 \qquad R_3 \rightarrow R_3 - 3R_1$$

We will then obtain the following reduced system, in matrix form:

$$\begin{bmatrix} 1 & -2 & 3 & 1 \\ 0 & 3 & -4 & -4 \\ 0 & 7 & -7 & -6 \end{bmatrix}$$

At this stage the second equation will now look like the second row of this matrix, that is,

$$3y - 4z = -4$$

We now use the y term in the second equation to eliminate the y term in the third equation. The row operation which will do this is

$$R_3 \rightarrow R_3 - (7/3)R_2$$

The resulting matrix is,

$$\begin{bmatrix} 1 & -2 & 3 & 1 \\ 0 & 3 & -4 & -4 \\ 0 & 0 & 7/3 & 10/3 \end{bmatrix}$$

At this stage we can stop reducing because we can start interpreting what we have arrived at.

For example the third row of this reduced augmented matrix is in reality the third equation in the form

$(7/3)z = 10/3$

Hence,

$z = 10/7$

Now we introduce the idea of back substitution. Using the value of z, just obtained, we back substitute it into the second equation, which now looks as

$3y - 4z = -4$

This gives us a value for y, since

$3y - 4(10/7) = -4$

and, hence,

$y = 4/7$

Finally we use the first equation, which in fact has not been altered throughout this elimination, and substitute the values of y and z already obtained to give us a value for x.

If you complete this exercise you should find that $x = -15/7$. Thus the unique solution is $x = -15/7$, $y = 4/7$, and $z = 10/7$.

You will recall that we stopped reducing the matrices further once we observed that we could find a value for z and thus back substitute to obtain the value of y and then, finally, the value of x.

The obvious question to ask is what happens if we continue to reduce the matrix to the bitter end. After all, DERIVE actually does that unless we use the functions SCALE_ELEMENT and SUBTRACT_ELEMENTS, which we introduced at the end of Chapter 4.

Let us continue from where we left off. We had the following matrix:

$$\begin{bmatrix} 1 & -2 & 3 & 1 \\ 0 & 3 & -4 & -4 \\ 0 & 0 & 7/3 & 10/3 \end{bmatrix}$$

The next thing we can do is to make the coefficient of y in the second equation equal to 1 by dividing the second equation throughout by 3. This is equivalent to the row operation

$R_2 \rightarrow R_2 / 3$

We then obtain the augmented matrix

$$\begin{bmatrix} 1 & -2 & 3 & 1 \\ 0 & 1 & -4/3 & -4/3 \\ 0 & 0 & 7/3 & 10/3 \end{bmatrix}$$

We now eliminate the y term from the first equation by applying the row operation

$R_1 \rightarrow R_1 + 2R_2$

The resulting matrix is

$$\begin{bmatrix} 1 & 0 & 1/3 & -5/3 \\ 0 & 1 & -4/3 & -4/3 \\ 0 & 0 & 7/3 & 10/3 \end{bmatrix}$$

Next we make the coefficient of the z term in row 3 equal to 1 by multiplying that row throughout by 3/7. The equivalent row operation is

$R_3 \rightarrow (3/7)R_3$

The resulting matrix is

$$\begin{bmatrix} 1 & 0 & 1/3 & -5/3 \\ 0 & 1 & -4/3 & -4/3 \\ 0 & 0 & 1 & 10/7 \end{bmatrix}$$

Notice that in this matrix we can read off the z value directly from row 3.

Finally we eliminate the z terms from both the first and second equations by using the z term of the third equation. The two row operations which produce these results are

$R_1 \rightarrow R_1 - (1/3)R_3$ and $R_2 \rightarrow R_2 + (4/3)R_3$

The resulting matrix should be as follows:

$$\begin{bmatrix} 1 & 0 & 0 & -15/7 \\ 0 & 1 & 0 & 4/7 \\ 0 & 0 & 1 & 10/7 \end{bmatrix}$$

From this system of equations we simply read the results, that once again $x = -15/7$, $y = 4/7$ and $z = 10/7$, as we did previously.

If we now compare the above ideas with work from preceding chapters we see that the augmented matrix has been row-reduced to bring it into reduced echelon form. We also observe that we started with $(A|\mathbf{k})$ and ended with $(I|\mathbf{x})$.

Exercise 5C

Use the Gaussian elimination row-reduction method to solve the following systems of equations:

(i) $2x - 3y + z = -1$
          $x + y + 3z = 12$
          $-x - 2y + z = -2$
(ii) $x + y + z = 0$
          $4x - y + 2z = 3$
          $x - 3y - z = 2$

DERIVE ACTIVITY 5B

Taking the example above use DERIVE, instead of doing all the calculations and manipulations by hand, to verify each of the steps we have made.

Many books present a reduction process called Gauss–Jordan elimination that avoids back substitution but this has disadvantages for large systems, for example ten equations in ten unknowns. The inverse matrix method of solving an $n \times n$ system of equations takes about $n^3 + n^2$ operations, whereas in Gaussian elimination the number of operations is about $n^3/3 + n^2 - n/3$. Thus **for large n** we have approximately n^3 versus $n^3/3$ operations; hence the Gaussian elimination procedure is used.

Checking

So far we have not looked at any intermediate checking of our work in the Gauss procedure. If we used DERIVE we wouldn't need such a check. If, however, we are doing row reduction by hand it is wise to have some check on whether our arithmetic is correct.

A simple checking procedure is given by the so-called row sum check. In this check we add a further column to our already augmented matrix and in it we put the sums of the rows one-by-one as we will see in the following example. Any operation we perform on the rows we also perform on the sum-check entries. At each stage in our reduction the row sum-check entries must also be correct.

By adding the entries of the relevant row, we see whether or not the sum-check column is correct. If the sum-check column of a particular row does not correspond to our addition of the elements of that row then we locate the error in that row by checking our arithmetic.

Take again the system of equations we used above:

$$\begin{aligned} x - 2y + 3z &= 1 \\ 2x - y + 2z &= -2 \\ 3x + y + 2z &= -3 \end{aligned}$$

Below we give the original augmented matrix, augmented by the row sum-check column.

$$\left[\begin{array}{ccc|c|c} 1 & -2 & 3 & 1 & 3 \\ 2 & -1 & 2 & -2 & 1 \\ 3 & 1 & 2 & -3 & 3 \end{array}\right]$$

The final column is obtained by adding the respective rows:

Row 1: $1 - 2 + 3 + 1 = 3$
Row 2: $2 - 1 + 2 - 2 = 1$
Row 3: $3 + 1 + 2 - 3 = 3$

Now if we start to row-reduce using the two row operations as we did earlier, namely,

$$R_2 \rightarrow R_2 - 2R_1 \qquad R_3 \rightarrow R_3 - 3R_1$$

we obtain the matrix

$$\begin{bmatrix} 1 & -2 & 3 & 1 & 3 \\ 0 & 3 & -4 & -4 & -5 \\ 0 & 7 & -7 & -6 & -6 \end{bmatrix}$$

We should now check that we have done the reduction correctly. In this case we have since, for example, in row 2, $0 + 3 - 4 - 4 = -5$.

Reducing further, making a deliberate mistake, while using the row operation

$$R_3 \rightarrow R_3 - (7/3)R_2$$

gives us the matrix shown below.

Identify the mistake and correct the offending entry:

$$\begin{bmatrix} 1 & -2 & 3 & 1 & 3 \\ 0 & 3 & -4 & -4 & -5 \\ 0 & 0 & 7/3 & 12/3 & 17/3 \end{bmatrix}$$

It should be clear that the third row is incorrect since

$$7/3 + 12/3 \neq 17/3$$

The correct entry can be found in our earlier work in this chapter.

Note that the unique solution will again come from this matrix by interpreting the equations obtained after the reductions.

Exercise 5D

1. For the problems in Exercise 5C show that the row sum is correct at each stage.
2. Use Gaussian elimination with row sum checks to solve the equations

$$\begin{aligned} 2x - 3y + z &= 4 \\ x + y - z &= 1 \\ 3x - 2y + 4z &= 0 \end{aligned}$$

5.5   PIVOTAL CONDENSATION

In most of the examples we have seen so far the coefficients have been exact and thus error-free. In practice the systems of equations have coefficients which are rounded-off or approximate in some way. Hence if we use such coefficients in our reduction method we could magnify the errors and obtain solutions which are totally unreliable, if not wrong!

In this section we will attempt to minimize this effect by a process known as *pivoting*. The method is reasonably straightforward and we will again proceed stage-by-stage, using an example to illustrate the procedure. We will use the same system of equations that we have used twice previously in this chapter:

$$x - 2y + 3z = 1$$
$$2x - y + 2z = -2$$
$$3x + y + 2z = -3$$

For the equations above look down the first column and find the largest coefficient, in modulus, that is, ignoring the sign of the term. It is 3 so this is the pivot and we would interchange rows 1 and 3, that is, $R_1 \leftrightarrow R_3$.

Definition

The *pivot* is the largest such coefficient. If the pivot is **not** in the first row, then interchange the first row and the pivot row.

Now the new pivot row is divided throughout by the pivot to obtain a coefficient of 1 for the first variable. We now have the system (with row sum checks)

$$\begin{bmatrix} 1 & 1/3 & 2/3 & -1 & 1 \\ 2 & -1 & 2 & -2 & 1 \\ 1 & -2 & 3 & 1 & 3 \end{bmatrix}$$

Now we use the pivot row to eliminate the 2 and 1 in the first column **below** the pivot. This means that we eliminate x from the second equation and the new third equation. After doing the necessary row operations already listed earlier in this chapter, we arrive at the matrix

$$
\begin{bmatrix}
1 & 1/3 & 2/3 & -1 & 1 \\
0 & -5/3 & 2/3 & 0 & -1 \\
0 & -7/3 & 7/3 & 2 & 2
\end{bmatrix}
$$

Now we continue the procedure by looking for the pivot in the second column, in modulus. In our case $-7/3$ is, in modulus, the largest value. Thus we must interchange row 2 and row 3, $R_2 \leftrightarrow R_3$. Note that from now on row 1 remains unaltered.

Next we divide the new pivotal row by $-7/3$ to obtain 1 as the coefficient for the y term in this pivotal row. We then have the system of equations in matrix form as follows:

$$
\begin{bmatrix}
1 & 1/3 & 2/3 & -1 & 1 \\
0 & 1 & -1 & -6/7 & -6/7 \\
0 & -5/3 & 2/3 & 0 & -1
\end{bmatrix}
$$

We continue as in the first stage by now using our new pivotal row to eliminate the $-5/3$, the element **below** the pivot. Hence we obtain the final matrix

$$
\begin{bmatrix}
1 & 1/3 & 2/3 & -1 & 1 \\
0 & 1 & -1 & -6/7 & -6/7 \\
0 & 0 & -1 & -10/7 & -17/7
\end{bmatrix}
$$

From this matrix we work out z as before and then back substitute to obtain y and x. As before, we obtain the unique solution

$$z = 10/7, \quad y = 4/7 \quad \text{and} \quad x = -15/7$$

Computers execute these algorithmic steps in solving systems of equations.

Exercise 5E

Use Gaussian elimination with row sum checks and pivoting to solve the following sets of equations:

(i) $\quad 2x - 3y + z = 4$
$\quad\quad x + y - z = 1$
$\quad\quad 3x - 2y + 4z = 0$

(ii) $\quad x - 2y + 3z = 4$
$\quad\quad 10x - 3y + 5z = 0$
$\quad\quad 3x + y - 2z = 1$

5.6   TABULAR METHOD FOR GAUSSIAN ELIMINATION

A tabular method for Gaussian elimination can compact the layout of your solutions, as the following example shows. You may well have thought in earlier work that once a row has been used and from that point on stays the same, it seems a duplication of work to rewrite it several times.

Example 5C

Solve the system of equations given below using row sum checks but **not** using pivotal condensation, as the coefficients are exact:

$$x_1 + x_2 - x_3 = -1$$
$$2x_1 - 3x_2 + 2x_3 = 4$$
$$3x_1 + 2x_2 - 4x_3 = -9$$

Solution

Tabular solution:

<u>1</u>	1	−1	−1	0
2	−3	2	4	5
3	2	− 4	−9	− 8
	<u>−5</u>	4	6	5
	−1	−1	− 6	− 8
		<u>−9/5</u>	−36/5	−9

We will now explain where this tableau has come from and how the values of x_1, x_2 and x_3 are obtained from it.

The first three rows in the tableau are the matrix of coefficients, augmented by the constants on the right-hand sides of the equations. The numbers after the vertical bar in these rows are the row sums. The number underlined in the first row is the element we use to eliminate the numbers underneath it. This means that the x_1 in the first equation is used to eliminate $2x_1$ and $3x_1$ in the second and third equations.

A horizontal bar is now drawn across the tableau. Immediately underneath this bar are the new second and third equations obtained from the elimination indicated above, together with their row sum checks after the vertical bar.

We underline the -5 in the second equation to indicate this is the element that we will use to eliminate the entries underneath it, in this case -1.

We draw a horizontal line underneath these new second and third equations and write underneath that line the result of eliminating the x_2 term in the third equation using the -5 indicated. Now we underline the $-9/5$.

We finally concentrate our attention on the three rows in which an element is underlined. Converting these back to full equations we have the following system:

$$
\begin{aligned}
x_1 + x_2 - \quad x_3 &= \quad -1 \\
-5x_2 + \quad 4x_3 &= \quad 6 \\
-(9/5)x_3 &= -36/5
\end{aligned}
$$

Now we can use the third equation to calculate the solution for $x_3 = 4$ and then back substitute this value into the second equation to obtain $x_2 = 2$ and finally back substitute both these values into the first equation to obtain $x_1 = 1$. Hence the unique solution in this case is $(1, 2, 4)$, or, in full, $x_1 = 1$, $x_2 = 2$ and $x_3 = 4$.

Although we have not used pivotal condensation because the coefficients are exact, the underlined numbers are still referred to as pivots, which are used to eliminate entries below them.

If we have a different system of equations with inexact coefficients, for example decimals rounded to two places, then we would use pivotal condensation to solve that system and underline the true pivots and rearrange the equations if necessary, as described in Section 5.5 above. Such a system will be given in tabular form in Section 5.8 below.

Exercise 5F

Rewrite the solution of the equations given in Exercise 5E using this tabular representation.

5.7   ILL-CONDITIONING

We now consider systems of equations which can produce 'answers' which vary widely because of rounding errors that can occur in calculating the coefficients.   Such systems have to be solved carefully in applications. We may be giving answers in good faith which are hugely inaccurate!

Definition

A system of equations is said to be *ill-conditioned* when small changes in some (or all) of the coefficients (and/or constants) produces large changes in the solutions.
   As an example of what can occur, consider the equations

$$x_1 + \quad x_2 = 1$$
$$x_1 + 1.01x_2 = 2$$

Take a moment to check that $x_1 = -99$ and $x_2 = 100$ is the unique solution.
   BUT what if the value 1.01 in the coefficient of x_2 in the second equation is not exact?
   Consider the system with one slight variation. Make the coefficient of x_2 in the second equation 1.02. What is the solution to the new system?
   By subtracting the first equation from the new second equation we would have

$$0.02x_2 = 1$$

From which

$$x_2 = 50$$

and hence, by substituting in the first equation we obtain

$$x_1 = -49$$

We hope this simple example gives you an idea of the problem. The solution to the second system is very different from the solution of the first system. Bearing in mind that in practical problems the coefficients are often obtained experimentally (and are thus subject to error), it can be appreciated that ill-conditioning can pose a serious threat to the validity of any solution obtained. In practice, we often have a rough idea of the magnitude of the solutions so we can generate solutions for different sets of equations with approximately the same coefficients.
   We will now indicate how we might try to be aware in a particular problem whether or not ill-conditioning is present. The clue lies in trying to solve $A\mathbf{x} = \mathbf{k}$ as

$\mathbf{x} = A^{-1}\mathbf{k}$, a result we have seen previously and indeed used to good effect when the solution is unique. For the problem above, $A\mathbf{x} = \mathbf{k}$ is equivalent to

$$\begin{bmatrix} 1 & 1 \\ 1 & 1.01 \end{bmatrix} \cdot \begin{bmatrix} x_1 \\ x_2 \end{bmatrix} = \begin{bmatrix} 1 \\ 2 \end{bmatrix}$$

Observe in this case that DET(A) is approximately equal to zero. In fact DET(A) = 0.01. What might this mean about the inverse of A? Before we answer this directly look again at the system of equations we are trying to solve, $A\mathbf{x} = \mathbf{k}$.

Assume that the coefficients are exact and that the entries in $\mathbf{k}$ are also exact. Through some inaccurate work the problem we actually solve is a perturbation of the original, that is, a small change in one or two coefficients. Hence we are not solving

$$A\mathbf{x} = \mathbf{k}$$

but, rather,

$$(A + \delta A) \cdot (\mathbf{x} + \delta \mathbf{x}) = \mathbf{k} + \delta \mathbf{k}$$

where δA, $\delta \mathbf{x}$ and $\delta \mathbf{k}$ are so called perturbation matrices. In simple terms these three matrices represent the inaccuracies. Let us now use the rules of matrices to manipulate this equation.

By expanding the parentheses and remembering that matrices do not commute, we have

$$A \cdot \mathbf{x} + A \cdot \delta \mathbf{x} + \delta A \cdot \mathbf{x} + \delta A \cdot \delta \mathbf{x} = \mathbf{k} + \delta \mathbf{k}$$

which simplifies, since $A\mathbf{x} = \mathbf{k}$, to

$$A \cdot \delta \mathbf{x} + \delta A \cdot \mathbf{x} + \delta A \cdot \delta \mathbf{x} = \delta \mathbf{k}$$

and further to

$$(A + \delta A) \cdot \delta \mathbf{x} = \delta \mathbf{k} - \delta A \cdot \mathbf{x}$$

In practice δA is not known, otherwise we could correct our mistakes or inaccuracies. To a **first** approximation,

$$\delta \mathbf{x} = A^{-1} \cdot (\delta \mathbf{k} - \delta A \cdot \mathbf{x})$$

Hence **if** elements of A^{-1} are large relative to those of $\delta \mathbf{k}$ and $\delta A \cdot \mathbf{x}$, then it is possible for $\delta \mathbf{x}$ to consist of large elements.

In the problem above we have

$$A^{-1} = \frac{1}{0.01}\begin{bmatrix} 1.01 & -1 \\ -1 & 1 \end{bmatrix}$$

$$= \begin{bmatrix} 101 & -100 \\ -100 & 100 \end{bmatrix}$$

From this we see that A^{-1} has relatively large elements.

A crude first approach to spotting ill-conditioning is therefore to consider DET(A) and see if it is very small in comparison to elements of adj(A). Then A^{-1} would contain large elements. This leads in Gaussian elimination to the condition that if the pivotal elements become necessarily small then ill-conditioning is likely.

There are classical examples of matrices which are ill-conditioned and these are known as *Hilbert matrices*, H_n, defined below.

Definition of Hilbert Matrices, H_n

The $n \times n$ Hilbert matrix, H_n, is given by the following formula. $H_n = (h_{ij})$ where $h_{ij} = 1/(i + j - 1)$ with i and j taking values from 1 to n.

H_2 would therefore be obtained by computing the elements h_{11}, h_{12}, h_{21}, and h_{22} using the formula above. These are given in full below:

$$h_{11} = 1/(1 + 1 - 1) = 1$$
$$h_{12} = 1/(1 + 2 - 1) = 1/2$$
$$h_{21} = 1/(2 + 1 - 1) = 1/2$$
$$h_{22} = 1/(2 + 2 - 1) = 1/3$$

Therefore,

$$H_2 = \begin{bmatrix} 1 & 1/2 \\ 1/2 & 1/3 \end{bmatrix}$$

Also from the formula for computing the entries of a Hilbert matrix it follows that $h_{ij} = h_{ji}$ and thus all Hilbert matrices are symmetrical.

DERIVE ACTIVITY 5C

(A)   Use DERIVE to calculate the solution of the matrix equation

$$H_3\mathbf{x} = (1\ \ 1\ \ 1)^T = \mathbf{k}$$

Written in full, we have to solve the matrix equation

$$\begin{bmatrix} 1 & 1/2 & 1/3 \\ 1/2 & 1/3 & 1/4 \\ 1/3 & 1/4 & 1/5 \end{bmatrix} \begin{bmatrix} x_1 \\ x_2 \\ x_3 \end{bmatrix} = \begin{bmatrix} 1 \\ 1 \\ 1 \end{bmatrix}$$

This has an exact solution obtainable by either Gaussian elimination or by inverse methods, or by using DERIVE. You should find that

$$x_1 = 3, \quad x_2 = -24 \quad \text{and} \quad x_3 = 30$$

Using DERIVE obtain H_3^{-1} and $DET(H_3)$.
        Verify that

$$H_3^{-1} = \begin{bmatrix} 9 & -36 & 30 \\ -36 & 192 & -180 \\ 30 & -180 & 180 \end{bmatrix}$$

and that

$$DET(H_3) = 1/2160$$

Hence,

$$\mathbf{x} = H_3^{-1}\mathbf{k}$$

gives the solution, $\mathbf{x}$ and thus the x_is, given above.

(B)   Now perturb the matrix H_3 slightly by giving the entries of H_3 in decimal form to three decimal places. We will denote this new matrix by H_3', which is given in full below:

$$H_3' = \begin{bmatrix} 1 & 0.500 & 0.333 \\ 0.500 & 0.333 & 0.250 \\ 0.333 & 0.250 & 0.200 \end{bmatrix}$$

If you use Gaussian elimination with a calculator with as full precision as possible, and back substitution, the following solutions are generated:

$$x_1 = 3.446\ 055\ 45, \quad x_2 = -26.273\ 519\ 19, \quad x_3 = 32.104\ 216\ 68$$

Try the calculations on your own calculator.

Input H_3' into DERIVE and solve the equation

$$\mathbf{x} = H_3'^{-1} \mathbf{k}$$

Stay in exact mode throughout.

$H_3'^{-1}$ should give you entries which are exact fractions, the denominator of each entry being 423 963.

When you ask DERIVE to solve for **x** you should obtain exact answers as follows:

$$x_1 = 487\ 000/141\ 321, \quad x_2 = -3\ 713\ 000/141\ 321,$$
$$x_3 = 4\ 537\ 000/141\ 321$$

Now ask for an approximation of these answers using **approX**. The results should be

$$x_1 = 3.446\ 05, \quad x_2 = -26.2735, \quad x_3 = 32.1042$$

To four decimal places these agree with the results you should have obtained from your calculator.

Let us now compare the solutions from the two matrices H and H' and the relative changes the slight perturbations bring about in these solutions. By relative changes we mean the difference between the two solutions, divided by the original value and then made into a percentage by multiplying by 100.

The symbol '$\cong$' used below means 'approximately equal to'. We have used the results obtained from DERIVE rather than the calculator.

H		H'
3	$\rightarrow$	3.446 05 with relative change $\cong$ 14.9%
−24	$\rightarrow$	−26.2735 with relative change $\cong$ 9.5%
30	$\rightarrow$	32.1042 with relative change $\cong$ 7.0%

As an example of how we obtained the percentages take the values obtained for x_1 by using H and H'.

The relative change is

$$\frac{(3.446\ 05 - 3)}{3} \times 100 \cong 14.9\%$$

We can see from this example that merely changing a fraction to an approximate decimal we have large differences in our answers, the relative change in x_1 being particularly large at almost 15%. As we said earlier in this section, H_3 is an example of an ill-conditioned matrix and this DERIVE ACTIVITY should convince you of this fact. It is also true that H_n in general is very ill-conditioned as n becomes larger and larger. H_n arises in regression problems for polynomials of degree $n - 1$.

(C)   Investigate the solution of the matrix equation

$$H_4 \mathbf{x} = (1 \ \ 1 \ \ 1 \ \ 1)^T$$

where

$$H_4 = \begin{bmatrix} 1 & \frac{1}{2} & \frac{1}{3} & \frac{1}{4} \\ \frac{1}{2} & \frac{1}{3} & \frac{1}{4} & \frac{1}{5} \\ \frac{1}{3} & \frac{1}{4} & \frac{1}{5} & \frac{1}{6} \\ \frac{1}{4} & \frac{1}{5} & \frac{1}{6} & \frac{1}{7} \end{bmatrix}$$

5.8   ITERATIVE METHOD

When a solution of the system of equations $A\mathbf{x} = \mathbf{k}$ is attempted, the computed solution $\mathbf{x}_c$ will not normally be equal to $\mathbf{x}$ as we have seen in the last section, for example. A more accurate result can be found by using *an iterative method*. We give such a method below.

We understand throughout this section that $\mathbf{x}$ is the solution while $\mathbf{x}_c$ is the calculated solution. The connection between $\mathbf{x}$ and $\mathbf{x}_c$ is some sort of error. We let $\mathbf{e}$ be the error vector in what follows and thus we have the equation

$$\mathbf{x} = \mathbf{x}_c + \mathbf{e}$$

which connects $\mathbf{x}$, $\mathbf{x}_c$ and $\mathbf{e}$.

Now let us return to the system of equations which we are trying to solve as accurately as possible:

$$A\mathbf{x} = \mathbf{k}$$

Hence,

$$A\mathbf{x} = A(\mathbf{x}_c + \mathbf{e}) = A\mathbf{x}_c + A\mathbf{e} = \mathbf{k}$$

From the last equality we have

$$A\mathbf{e} = \mathbf{k} - A\mathbf{x}_c$$

We put

$$\mathbf{r} = \mathbf{k} - A\mathbf{x}_c$$

where $\mathbf{r}$ is known as the *residual vector*.

It follows from these last two equations that a solution to $A\mathbf{e} = \mathbf{r}$ will give an approximation for $\mathbf{e}$. As this is the first approximation for the error vector we will label it $\mathbf{e}_1$. This then gives us an improved solution to our original problem, since

$$\mathbf{x} \cong \mathbf{x}_c + \mathbf{e}_1$$

In our iteration procedure we will call this improved solution $\mathbf{x}_1$. Thus,

$$\mathbf{x}_1 = \mathbf{x}_c + \mathbf{e}_1$$

The residual vector associated with this first approximate solution is thus equal to

$$\mathbf{r} = \mathbf{k} - A \cdot \mathbf{x}_1$$

This residual vector can be used to further improve the approximation for $\mathbf{x}$ and the process is iterated until the estimated error vector makes no significant difference to the accuracy of the computed solution. This technique is used in all good stable algorithms; however, a word of caution is necessary.

It is essential that the residual vector **r** be calculated very accurately. To achieve this on a computer, precision arithmetic appropriate to the language of the program would be used. DERIVE can be used to good effect on this problem of accuracy.

Consider the following system of equations:

$$1.1x_1 + 4.9x_2 - 2.9x_3 = 18.3$$
$$-1.9x_1 + 2.1x_2 + 6.9x_3 = 19.4$$
$$9.1x_1 - 1.9x_2 + 1.2x_3 = 47.8$$

We attempt a solution using Gaussian elimination with row pivoting, and recording answers to three significant figures and using multipliers corrected to three significant figures. No sum checks will be used. We will explain the idea of multipliers below.

We will do the same exercise using DERIVE after we have seen what is going on by completing this exercise by hand, with the aid of a calculator. At this stage you will need to recall the tableau method introduced in Section 5.6, as well as the idea of pivotal condensation introduced in Section 5.5.

The process begins by looking for the first pivot. This is in the third equation and has the value '9.1'. Thus we need to interchange the first equation and third to start our tableau. The pivots are again underlined for convenience in the tableau given below.

The column marked m gives the relevant *multiplier* of the pivot to eliminate the entries below it. An example, in the tableau shown below, will be done to illustrate this. The tableau is as follows:

m				
	$\underline{9.1}$	-1.9	1.2	47.8
-0.209	-1.9	2.1	6.9	19.4
0.121	1.1	4.9	-2.9	18.3
		1.70	7.15	29.4
		5.13	-3.05	12.5
		$\underline{5.13}$	-3.05	12.5
0.331		1.70	7.15	29.4
			$\underline{8.16}$	25.3

Hence $R_2 \leftrightarrow R_3$

At this stage we will explain exactly where we have got this tableau from. The first three lines hold the coefficients and constants from the three equations, but the first and third equations have been interchanged because of the pivotal value.

We use the '9.1' to eliminate the '–1.9' and '1.1' which lie below it. To eliminate –1.9 using 9.1 we would need a multiplier of – 0.209 because

$$(- 0.209)(9.1) = -1.9019$$

Remember, we are working to three significant figures for the multipliers. We place this '– 0.209' in the m column for reference.

Now we use it three more times to obtain a new second equation, which is written below the horizontal line. The calculations are listed below:

$$2.1 - (- 0.209)(-1.9) = 1.70$$
$$6.9 - (- 0.209)(1.2) = 7.15$$
$$19.4 - (- 0.209)(47.8) = 29.4$$

Similarly, we find that the multiplier 0.121 will be needed to eliminate the 1.1, since $(0.121)(9.1) = 1.1011$. Thus we write this multiplier again in the m column for reference. Having used it we obtain the fifth line of the tableau and a new third equation.

At this stage we look for the second pivot in the second and third equations below the first horizontal line. We observe that the pivot will be 5.13 and it is in the third equation, therefore we must interchange the second and third equations or R_2 and R_3, as indicated.

Under the second horizontal line we now write the pivotal equation and then the remaining equation. The multiplier we need to use in the final reduction is 0.331, which will eliminate the 1.70 sitting below the pivot, 5.13, since

$$(0.331)(5.13) = 1.698\ 03$$

Again we write the multiplier in the m column for reference and use it to obtain a new third equation which we write under the third horizontal line. There is no choice for the final pivot.

We are now ready to consider the three pivotal equations. These are the ones in which an element is underlined. We start with the third equation, which is

$$8.16x_3 = 25.3$$

This simplifies to

$$x_3 = 3.10$$

Hence we obtain by back substitution a value for x_2 from the second equation, which is

$$5.13x_2 - 3.05x_3 = 12.5$$

Solving for x_2 gives the value 4.28, to three significant figures.
        Finally, we obtain a value from the first equation, which is

$$9.1x_1 - 1.9x_2 + 1.2x_3 = 47.8$$

This gives x_1 as 5.74.
        We can write the total solution in vector form as follows:

$$\mathbf{x}_c^T = (5.74 \quad 4.28 \quad 3.10)$$

Having explained the tableau in some detail we move on to take this computed solution, which is an approximation to the 'actual solution', and consider the residual vector. We evaluate the residual vector from the equation

$$\mathbf{r} = \mathbf{k} - A.\mathbf{x}_c$$

as follows:

$$\mathbf{r} = \begin{bmatrix} 18.3 \\ 19.4 \\ 47.8 \end{bmatrix} - \begin{bmatrix} 1.1 & 4.9 & -2.9 \\ -1.9 & 2.1 & 6.9 \\ 9.1 & -1.9 & 1.2 \end{bmatrix} \cdot \begin{bmatrix} 5.74 \\ 4.28 \\ 3.10 \end{bmatrix} = \begin{bmatrix} 0.004 \\ -0.072 \\ -0.022 \end{bmatrix}$$

We continue to follow the theory outlined at the beginning of this section.
        The solution of $A \cdot \mathbf{e} = \mathbf{r}$ is now attempted using row pivoting and three significant figure accuracy as before. We let $\mathbf{e} = [e_1 \ e_2 \ e_3]$, the error vector we wish to calculate. The full equations look as follows:

$$1.1e_1 + 4.9e_2 - 2.9e_3 = 0.004$$
$$-1.9e_1 + 2.1e_2 + 6.9e_3 = -0.072$$
$$9.1e_1 - 1.9e_2 + 1.2e_3 = -0.022$$

We solve these again in tableau form and notice, before we write the tableau down, that the first pivot is still in the third equation. Thus we interchange the first and third equations or rows, as we did before. We obtain the following tableau:

m		A		r
	9.1	−1.9	1.2	− 0.022
− 0.209	−1.9	2.1	6.9	− 0.072
0.121	1.1	4.9	−2.9	0.004
		1.70	7.15	− 0.0766
		5.13	−3.05	0.006 66
0.331			8.16	− 0.0788

We use the same reasoning as before to obtain the pivots and the final solution for **e**. From the final equation we have

$$8.16e_3 = -0.0788$$

which gives

$$e_3 = -0.009\ 65$$

Back substitution yields the full solution:

$$\mathbf{e}_c^T = [-0.002\ 07 \quad -0.004\ 45 \quad -0.009\ 65]$$

Note again that this is the computed error term **e** to three significant figures and so we have labelled it $\mathbf{e}_c$. Thus the improved solution $\mathbf{x}_c + \mathbf{e}_c$ is given by

$$\begin{bmatrix} 5.74 \\ 4.28 \\ 3.10 \end{bmatrix} + \begin{bmatrix} -0.002\ 07 \\ -0.004\ 45 \\ -0.009\ 65 \end{bmatrix} = \begin{bmatrix} 5.737\ 93 \\ 4.275\ 55 \\ 3.090\ 35 \end{bmatrix}$$

Thus to three significant figures the improved solution becomes

$$[5.74 \quad 4.28 \quad 3.09]$$

We now iterate again and find the new residual vector from the equation

$$\mathbf{r} = \mathbf{k} - A \cdot (\mathbf{x}_c + \mathbf{e}_c)$$

The new residual vector r^T is now found to be

$$[0.025 \quad -0.003 \quad -0.01]$$

Exercise 5G

(i)   Compute the next estimate for the error and continue the process until x_c is correct to three significant figures.

(ii)   Carry out the iterative method for the following systems of equations:

(a) $1.3x_1 - 4.2x_2 = 16.1$

$11.1x_1 + 1.3x_2 = \quad 5.7$

(b) $0.3x_1 - \quad 4.1x_2 + 2.7x_3 = 11.2$

$4.6x_1 + \quad 2.5x_2 - 3.1x_3 = \quad 9.7$

$5.1x_1 - 11.2x_2 + 3.6x_3 = 13.4$

DERIVE ACTIVITY 5D

As we said earlier we can do all this work using DERIVE in a fraction of the time, as this activity will now show.

(A)   Input the matrix of coefficients, given below, on line 1:

$$\begin{bmatrix} 1.1 & 4.9 & -2.9 \\ -1.9 & 2.1 & 6.9 \\ 9.1 & -1.9 & 1.2 \end{bmatrix}$$

On line 2 input the matrix

$$k = \begin{bmatrix} 18.3 \\ 19.4 \\ 47.8 \end{bmatrix}$$

On line 3 **Author** the inverse of line 1 and then on line 4 **Simplify** line #3.

As you are in exact mode your matrix on line 4 will have fractions for its entries with denominator, 190493.

On line 5 **Author**

#4.#2

This in reality is

$$A^{-1}\mathbf{k}$$

On line 6 **Simplify** the expression from line 5. This is the solution vector **x**.
 Finally on line 7 **approX** line #6. Your 3 × 1 matrix in line 7 should be

$$\begin{bmatrix} 5.737\,92 \\ 4.275\,56 \\ 3.090\,34 \end{bmatrix}$$

Now compare these entries with the solution you obtained by hand above. To three significant figures they agree exactly with the improved solution found previously.

The result was achieved in seven lines of DERIVE, showing its power to cut down on working once we have understood the theory. You are recommended to use DERIVE to repeat the steps in the text above to see at each step the accuracy you can expect from the package for such things as the residual vector and the error vector.
 Note that if a solution is accurate then it is essential that the residual vector be small. However, a small residual vector does not necessarily imply an accurate solution, as illustrated by Exercise 5H.

Exercise 5H

Consider the two equations given below:

$$0.89x_1 + 0.53x_2 = 0.36$$
$$0.47x_1 + 0.28x_2 = 0.19$$

Working with three significant figures obtain the approximate solution

$$x_c^T = [0.702 \ -0.500]$$

and the corresponding residual vector

$$\mathbf{r} = \begin{bmatrix} 0.000\,22 \\ 0.000\,06 \end{bmatrix}$$

In the exercise, you might, if too hasty, conclude that the solution is accurate as the two elements in the residual vector **r** are small.

However, the exact solution is

$$\mathbf{x}^T = [1 \; -1]$$

as can be checked by direct substitution.

On this word of caution we conclude this chapter.

Consolidation Exercise

1. Use Gaussian elimination with pivotal condensation and row sum checks to solve the following system of equations exactly. Then check your results using DERIVE:

$$2x_2 + x_3 = 3$$
$$-2x_1 + 2x_2 + x_3 = 4$$
$$x_1 - x_2 + x_3 = 1$$

2. Use Gaussian elimination to show that the system of equations given below has no solution:

$$x_1 - 2x_2 + x_3 = 3$$
$$-x_1 + 4x_2 = -1$$
$$2x_1 + 4x_3 = 12$$

3. Consider the system of equations given below:

$$2x_1 + 3x_2 - x_3 = 5$$
$$4x_1 + 4x_2 - 3x_3 = 3$$
$$2x_1 - 3x_2 + x_3 = -1$$

   (i)   Solve them using Cramer's rule.
   (ii)   Solve them using the matrix method.
   (iii) Solve them using Gaussian elimination
   (iv)   Check your results using DERIVE.

4. Let A be a 2×3 matrix with rank 2.

   Let **x** be a 3×1 matrix and **b** be a 2×1 matrix.

   Show that the 2×3 system of equations, $A\mathbf{x} = \mathbf{b}$, is consistent for every choice of **b** in $\Re^2$.

5.   Recall that homogeneous equations always have at least one solution. Consider the systems of homogeneous equations given below and solve them using DERIVE and its row reduce facility:

(i) $2x_2 + x_3 = 0$
$$x_1 - 3x_2 + 2x_3 = 0$$
$$2x_1 + x_2 - x_3 = 0$$

(ii) $x_1 - 2x_2 - x_3 = 0$
$$x_2 + x_3 = 0$$
$$x_1 + 3x_2 + 4x_3 = 0$$

6.   Show that the following system of equations is ill-conditioned if we assume that rounding has been done to three significant figures:

$$x_1 + x_2 = 50$$
$$x_1 + 1.026x_2 = 20$$

What is the relative error (to the nearest whole number) in x_1 and x_2 which comes from the rounding process?
What is the exact solution?

6

Vector Spaces and Linear Transformations

In this chapter we introduce the idea of vector spaces using some of the ideas introduced in Chapters 1 and 3. The objects we are studying are vectors and can be thought of as column or row matrices. Furthermore, they can be thought of as ordered *n*-tuples, as introduced at the beginning of Chapter 3. Before proceeding you are advised to revise the early part of Chapter 3 and the properties of matrices found in Chapter 1.

6.1   VECTOR SPACES

From the fact that vectors can be considered as matrices, properties which matrices have apply equally to vectors. Let $\mathbf{u}$, $\mathbf{v}$ and $\mathbf{w}$ be three vectors. They can have any number of elements, but it is often easier to think of vectors with two or three elements in examples. Let $\mathbf{u} = [1 \quad 2 \quad 3]$, $\mathbf{v} = [-2 \quad 0 \quad -1]$, $\mathbf{w} = [0 \quad -3 \quad 2]$ and $\mathbf{0} = [0 \quad 0 \quad 0]$ (the zero or null vector) be vectors belonging to $\mathfrak{R}^3$; then we can write down the vectors formed from these:

(i) $\mathbf{v} + \mathbf{u}$
(ii) $\mathbf{u} + \mathbf{v}$
(iii) $(\mathbf{u} + \mathbf{v}) + \mathbf{w}$
(iv) $\mathbf{u} + (\mathbf{v} + \mathbf{w})$
(v) $2\mathbf{u} + 5\mathbf{v} - \mathbf{w}$
(vi) $\mathbf{v} + -\mathbf{v}$
(vii) $\mathbf{w} + \mathbf{0}$

The results of these examples would be

(i) $[-2 + 1 \quad 0 + 2 \quad -1 + 3] = [-1 \quad 2 \quad 2]$
(ii) $[1 + -2 \quad 2 + 0 \quad 3 + -1] = [-1 \quad 2 \quad 2]$
(iii) $([1 \quad 2 \quad 3] + [-2 \quad 0 \quad -1]) + [0 \quad -3 \quad 2] = [-1 \quad 2 \quad 2] + [0 \quad -3 \quad 2] = [-1 \quad -1 \quad 4]$

(iv) $[1 \ 2 \ 3] + ([-2 \ 0 \ -1] + [0 \ -3 \ 2])$

$= [1 \ 2 \ 3] + [-2 \ -3 \ 1]$

$= [-1 \ -1 \ 4]$

(v) $2[1 \ 2 \ 3] + 5[-2 \ 0 \ -1] - [0 \ -3 \ 2]$

$= [2 \ 4 \ 6] + [-10 \ 0 \ -5] - [0 \ -3 \ 2]$

$= [-8 \ 7 \ -1]$

(vi) $[-2 \ 0 \ -1] + -[-2 \ 0 \ -1]$

$= [-2 \ 0 \ -1] + [2 \ 0 \ 1]$

$= [0 \ 0 \ 0] = \mathbf{0}$

(vii) $[0 \ -3 \ 2] + [0 \ 0 \ 0]$

$= [0 + 0 \ -3 + 0 \ 2 + 0]$

$= [0 \ -3 \ 2] = \mathbf{w}$

From these examples, we observe some results which can be generalized. These results, together with a few others, will form a set of axioms for a vector space.

First we observe that vectors commute, our example being that

$$\mathbf{v} + \mathbf{u} = \mathbf{u} + \mathbf{v}$$

Next we observe that vectors are associative under addition. In our example we saw that

$$(\mathbf{u} + \mathbf{v}) + \mathbf{w} = \mathbf{u} + (\mathbf{v} + \mathbf{w})$$

Also to each vector $\mathbf{v}$ there is an inverse vector under addition, $-\mathbf{v}$, such that

$$\mathbf{v} + -\mathbf{v} = \mathbf{0}$$

Finally in our observations we had the zero vector under addition leaving the vector $\mathbf{w}$ identical to what it was at the start:

$$\mathbf{w} + \mathbf{0} = \mathbf{w}$$

In this role, the zero vector is called the identity under addition.

We now give the definition of a vector space in general terms.

Let $V = \{v_1, v_2, v_3, \ldots, v_{k-2}, v_{k-1}, v_k\}$ be a set of vectors in $\Re^n$. Let F be a set of scalars, as, for example, the real or complex numbers. The elements of F will be denoted by Greek letters such as α, β, γ to distinguish them from the vectors. With these preliminaries we can now formally define a vector space V.

Definition

Let V be a non-empty set of vectors on which a law of addition is defined. Let F be a set of scalars. Further, scalar multiplication is well defined. This means that given α in F and v in V, αv is well defined.

A *vector space V over F*, is defined to be a non-empty set V of vectors together with a set F of scalars which obey the following axioms:

(i)   The set V is closed under vector addition, this means, given v_1 and v_2 in V, $v_1 + v_2$ is in V.

(ii)   The vectors in V are associative:

$$(v_i + v_j) + v_k = v_i + (v_j + v_k)$$

for any i, j, k.

(iii)   There is a vector in V, namely, the zero vector, 0, such that

$$v_i + 0 = v_i$$

for any i.

(iv)   For each vector v_i in V there is a vector $-v_i$ in V called its additive inverse, such that

$$v_i + -v_i = 0$$

(v)   The vectors in V are commutative:

$$v_i + v_j = v_j + v_i$$

for any i and j.

(vi)   The scalar multiple αv belongs to V for any α in F and any v in V.

(vii)   For any scalar α in F and any vectors v_i and v_j in V:

$$\alpha(v_i + v_j) = \alpha v_i + \alpha v_j$$

(viii) For any scalars α and β in F and any vector $\mathbf{v}_i$ in V:

$$(\alpha + \beta)\mathbf{v}_i = \alpha\mathbf{v}_i + \beta\mathbf{v}_i$$

(ix)   For any scalars α and β in F and any vector $\mathbf{v}_i$ in V:

$$(\alpha\beta)\mathbf{v}_i = \alpha(\beta\mathbf{v}_i)$$

(x)   For the unit scalar, 1, in F:

$$1\mathbf{v}_i = \mathbf{v}_i$$

for all i.

While these might look formidable they all follow from observations of the kind we listed earlier in this chapter.

The axioms split into two distinct kinds.

The first five axioms tell us how vectors behave under addition. If the reader is familiar with the idea of a group, the first five axioms can be summarized as follows: 'vectors form an abelian group under addition'. The last five axioms for a vector space are all concerned with how the scalars interact with the vectors.

From the fact that vectors associate it follows that you never need to worry about parentheses when adding vectors. However, for convenience in some calculations it is better to put them in when the expressions become a little complicated.

From the fact that vectors commute, we do not need to worry about the order in which we write vectors down.

From the fact that additive inverses are unique it follows that we can cancel vectors from both sides of an equation, that is, $\mathbf{v}_i + \mathbf{v}_j = \mathbf{v}_k + \mathbf{v}_j$ implies $\mathbf{v}_i = \mathbf{v}_k$. Furthermore, we can define subtraction of vectors by the equation

$$\mathbf{v}_i - \mathbf{v}_j = \mathbf{v}_i + -\mathbf{v}_j$$

From the axioms other results can be proved in vector spaces, as we will indicate in the exercise given below.

Exercise 6A

Let V be a vector space and F a set of scalars. Prove the following results using any of the axioms listed earlier.

(i)   For any scalar α in F:

$$\alpha 0 = 0$$

where **0** is the zero vector in V.
Hint. Use the third and seventh axioms.

(ii)   For the zero scalar, 0, in F and any vector $\mathbf{v}_i$ in V:

$$0\mathbf{v}_i = 0$$

Hint. Use the eighth axiom and the fact that $0 + 0 = 0$ in F.

(iii)   If $\alpha\mathbf{v}_i = 0$ then either $\alpha = 0$ or $\mathbf{v}_i = 0$
Hint. Use the property of F that if $\alpha \neq 0$ then there exists α^{-1}.

(iv)   For any scalar α in F and any vector $\mathbf{v}_i$ in V:

$$(-\alpha)\mathbf{v}_i = \alpha(-\mathbf{v}_i) = -\alpha\,\mathbf{v}_i$$

Although these results might seem 'obvious' they can give trouble if the reader is not familiar with using axioms to prove other results. Where do we start? Which axiom(s) do we use? We have given hints as to how to approach the first three.
We prove (iv) below.

Solution to (iv)

Begin with the inverse axiom and the result of the first part of this exercise, that is,

$$\mathbf{v}_i + -\mathbf{v}_i = 0$$

and

$$0 = \alpha 0$$

Then

$$0 = \alpha 0 = \alpha(\mathbf{v}_i + -\mathbf{v}_i) = \alpha\mathbf{v}_i + \alpha(-\mathbf{v}_i)$$

Hence we have

$$0 = \alpha\mathbf{v}_i + \alpha(-\mathbf{v}_i)$$

Now we add $-(\alpha\mathbf{v}_i)$ to both sides of this equation to obtain

$$-(\alpha\mathbf{v}_i) = \alpha(-\mathbf{v}_i) \tag{6.1}$$

This gives us part of our total result.

Now begin with $\alpha + -\alpha = 0$ a fact from F, and $0 = 0\mathbf{v}_i$ (from (ii) in this exercise). Hence,

$$0 = 0\mathbf{v}_i = (\alpha + -\alpha)\mathbf{v}_i = \alpha\mathbf{v}_i + (-\alpha)\mathbf{v}_i$$

from which the equation

$$0 = \alpha\mathbf{v}_i + (-\alpha)\mathbf{v}_i$$

can be written down.

As before, add $-(\alpha\mathbf{v}_i)$ to both sides of this equation to obtain the result

$$-(\alpha\mathbf{v}_i) = (-\alpha)\mathbf{v}_i \tag{6.2}$$

Finally, combine (6.1) and (6.2) to obtain the required result:

$$-(\alpha\mathbf{v}_i) = \alpha(-\mathbf{v}_i) = (-\alpha)\mathbf{v}_i$$

6.2   EXAMPLES OF VECTOR SPACES

To give an example of a vector space we need to state clearly what the set V contains and we need to state the set of scalars F that we are using. Then we must give the law of addition on V, that is, how do we add two elements of V: and we need to define scalar multiplication, that is, how to evaluate $\alpha\mathbf{v}$, and thus how the scalars interact with the vectors.

(i)   The first example is the most natural one in the sense that the objects we have for V are vectors in the traditional sense.

Recall at the beginning of Chapter 3 that we defined $\Re^2$ as a set of vectors with two real coordinates.

Let $V = \Re^2$ and let $F = \Re$, then V is a vector space over $\Re$.

Let $\mathbf{v}_1 = [x_1 \ y_1]$ and $\mathbf{v}_2 = [x_2 \ y_2]$ be two vectors in $V = \Re^2$

We define addition as follows:

$$\mathbf{v}_1 + \mathbf{v}_2 = [x_1 \ y_1] + [x_2 \ y_2] = [x_1 + x_2 \ y_1 + y_2]$$

Notice that the sum of $\mathbf{v}_1$ and $\mathbf{v}_2$ is contained in V.

We define scalar multiplication as follows:

$$\alpha\mathbf{v}_1 = \alpha[x_1 \ y_1] = [\alpha x_1 \ \alpha y_1]$$

With these definitions it should be a straightforward matter to check the axioms. We can generalize this example and show that $\mathfrak{R}^n$ is a vector space over $\mathfrak{R}$. We will use this example on several occasions later in this chapter. For a second example we consider matrices.

(ii)   Let V be the set of $m \times n$ matrices with entries from the set of scalars F. Here F could be the real or complex numbers. The operation of addition on V will be the ordinary addition of two $m \times n$ matrices as defined in Chapter 1. The operation of scalar multiplication will also be the one that was introduced in Chapter 1, namely,

$$\alpha(a_{ij}) = (\alpha a_{ij})$$

With these operations V will be a vector space over F. Check the axioms to verify this statement. Everything you need to complete this example is in Chapter 1.

For our third example we move away from traditional vectors and matrices.

(iii)   Consider expressions of the form

$$a_n x^n + a_{n-1} x^{n-1} + \ldots + a_2 x^2 + a_1 x + a_0$$

This is called a polynomial of degree n.

The coefficients, a_i, are taken from a set of scalars F. Again think of the real or complex numbers for these values. Now let V be the set of all polynomials of degree n (or less) with coefficients in F. V is a vector space over F if we define addition as follows: let

$$v_1 = a_n x^n + a_{n-1} x^{n-1} + \ldots + a_1 x + a_0$$

and

$$v_2 = b_n x^n + b_{n-1} x^{n-1} + \ldots + b_1 x + b_0$$

then

$$v_1 + v_2 = (a_n + b_n)x^n + (a_{n-1} + b_{n-1})x^{n-1} + \ldots + (a_1 + b_1)x + (a_0 + b_0)$$

In elementary algebra we would say that addition is defined as adding like terms.

We define scalar multiplication in a natural way as follows:

$$\alpha \mathbf{v}_1 = \alpha(a_n x^n + a_{n-1} x^{n-1} + \ldots + a_2 x^2 + a_1 x + a_0)$$

$$= \alpha a_n x^n + \alpha a_{n-1} x^{n-1} + \ldots + \alpha a_2 x^2 + \alpha a_1 x + \alpha a_0$$

With these two well-defined operations we have an example of a vector space. Again check the axioms for yourself.

We give a fourth example below which concerns particular functions.

(iv)　　Let V be the set of real valued functions of a real variable, that is, the functions have real numbers both for their domain and codomain. We assume that the reader is familiar with this idea either from using DERIVE or from the literature.

If in doubt about this idea, the reader is strongly recommended to refer to the first book in the present series, *Learning mathematics through DERIVE* by Berry, Graham and Watkins.

Now we return to this fourth example.

Let V be the set of real valued functions of a real variable. Let $F = \Re$, the set of real numbers.

Hence $V = \{f, g, \ldots\}$, that is, $f : \Re \rightarrow \Re$ and $g : \Re \rightarrow \Re$, and similarly for other members of V.

Given f and g in V we need to define $f + g$ and αf, where α is in $F = \Re$.

Define addition as follows:

$$(f + g)x = f(x) + g(x)$$

This is the usual addition of functions.

Define scalar multiplication as follows:

$$(\alpha f)x = \alpha(f(x))$$

Note that both $f + g$ and αf belong to V, as they should from our definition of a vector space.

With these operations V is a vector space over $\Re$. Again check the axioms for yourself.

There are many other examples in the literature, but these four will suffice for our purposes.

6.3   VECTOR SUBSPACES

Before we leave this section we introduce the idea of subspaces.

Definition

Consider a vector space V. A *subspace* of V is a collection of vectors from V which itself satisfies the axioms for a vector space.

Consider the third example given above. Here we had a vector space consisting of polynomials of degree n or less. If we take just the polynomials of the form $a'x^3$, where a' is a real number, we can check the axioms for a vector space and show that they are all satisfied. Do this now to your own satisfaction. Thus this set of polynomials is a subspace of all polynomials of degree less than or equal to n where $n \geq 3$. We will meet other examples of subspaces in what follows.

Criteria for a Subspace

Let V be a vector space and let U be a subset of V. To prove that U is a subspace of V, we only need to check three axioms, namely,

(i)   the zero vector of V is in U,
(ii)   if u_1 and u_2 are in U then $u_1 + u_2$ is in U, and
(iii)   if u is in U and α is any scalar, then αu is in U.

Another way of saying this is that the subset U must be closed under vector addition and under scalar multiplication, and the zero vector must be in U. It is a good working criterion to check the first axiom above, as, if we inspect U in a particular problem and find that the zero vector is not contained in it, we need go no further. We would declare that U is **not** a subspace of V.

The proof of the subspace criteria is not too complicated, as we will endeavour to show below.

Proof of Subspace Criteria

If U is a subspace of V, then it must satisfy all the axioms of a vector space by the definition of subspace. In particular, it must satisfy the three conditions given in the criteria as these are three of the axioms. Conversely, suppose that U satisfies the three conditions above; we now have to show that all the other axioms given in Section 6.1 are satisfied. First, the three conditions are in fact axioms (i), (iii) and (vi), so they are satisfied.

Second, axioms (ii), (v), (vii), (viii), (ix) and (x) are all 'inherited' from the vector space V in which U lies. For example, if vectors are associative in V, then they must stay associative in the 'smaller' set U. Therefore, we are left to show that axiom

(iv) is satisfied. In Exercise 6A we showed that $(-\alpha)\mathbf{v}_i = \alpha(\mathbf{v}_i) = -\alpha\,\mathbf{v}_i$ and this must remain true for elements $\mathbf{u}_i$ in U and α in F.

Also, axiom (x) says

$$1\mathbf{v}_i = \mathbf{v}_i$$

So, choosing $\alpha = 1$ in F and $\mathbf{u}_i$ in U, we have the following true statements:

$$(-1)\mathbf{u}_i = 1(-\mathbf{u}_i) = -1\mathbf{u}_i \quad \text{and} \quad 1\mathbf{u}_i = \mathbf{u}_i$$

and thus

$$(-1)\mathbf{u}_i = -\mathbf{u}_i$$

But by condition (iii) in the subspace criteria we have that if α is in F and $\mathbf{u}$ is in U then $\alpha\mathbf{u}$ is in U, from which it follows that $(-1)\mathbf{u}_i = -\mathbf{u}_i$ must belong to U, as required.

Hence axiom (iv) of the vector space axioms is satisfied and thus all the vector space axioms are satisfied by U.

The criteria given for vector subspaces prove very useful in practice, as we will see later in this chapter.

Exercise 6B

1. Suppose that $\mathbf{U} = \{[a,b,-a,-b]^T: a,b\in\Re\}$ is a subset of $\Re^4$. Show that $\mathbf{U}$ is a subspace of $\Re^4$.
2. For each of the following vector spaces V and subsets S, decide whether S is a subspace of V:
    (i) $V = \{[a,b]^T: a,b\in\Re\}$, $S = \{[a,b]^T: a+b=0\}$
    (ii) $V = \{[a,b,c]^T: a,b,c\in\Re\}$, $S = \{[a,b,c]^T: a+b-c=0\}$
    (iii) $V = \{[a,b]^T: a,b\in\Re\}$, $S = \{[a,b]^T: a^2-b^2=1\}$

6.4   SPANS, FRAMES AND BASES

Now we have introduced vector spaces and subspaces, we move on to consider certain important ideas which are associated with them. In Chapter 3 we introduced the idea of linear combinations of vectors and sets of vectors which were either independent or dependent. Before proceeding further you might like to revise those ideas.

A set of linearly independent vectors is also known as a *frame*. It is a good name for independent vectors as they are the building elements for further vectors. They are the frame on which we build more vectors by the process of linear combinations.

Consider for example, the two vectors [1 0] and [0 1] in $\Re^2$. If we take linear combinations of these two vectors we can in fact produce all the vectors in $\Re^2$, since

$$x[1 \ \ 0] + y[0 \ \ 1] = [x \ \ y]$$

We could introduce a third vector, say [1 1], to our set, which would now give us the set $V = \{[1 \ \ 0], [0 \ \ 1], [1 \ \ 1]\}$. This set is now a dependent set since a relationship can be written down, such as

$$[1 \ \ 0] + [0 \ \ 1] = [1 \ \ 1]$$

In fact any one of the three vectors is redundant!

Note. The set V is not a vector space for many reasons. One reason would be that the zero vector is not contained in V as it should be for a vector space.

Now the dependent set V gives us no more information than the original set with just the two vectors in it. No matter how many linear combinations we take with the three vectors we will not produce any more vectors than we did with the vectors [1 0] and [0 1], that is, **i** and **j**. This idea of generating more vectors from given vectors leads us into the concept of *a spanning* set or *a span*.

Consider the vector [1 0] on its own. What can this one vector generate or span? We can take linear combinations of it to produce multiples of it and that is the best we can do. So [1 0] spans the vectors $x[1 \ \ 0]$ where x is any real number.

Thinking of this in the plane we are spanning a line. Thinking of this on a graph we are spanning the x axis. Is the set of vectors $\{x[1 \ \ 0]\}$ a subspace of $\Re^2$?

Definition

Let $V = \{\mathbf{v}_1, \mathbf{v}_2, ..., \mathbf{v}_{k-1}, \mathbf{v}_k\}$ be a set of vectors. The *span* of V is the set of all vectors that can be generated from the vectors of V by taking linear combinations of the elements of V.

Example 6A

For each of the sets V given below, decide what the span of V is in each case. Recall that $\Re^n$ is a vector space for every natural number n. The set of vectors given in each case is taken from $\Re^n$ for a suitable n:

(i) $V = \{[0 \ \ 1]\}$ in $\Re^2$.
(ii) $V = \{[1 \ \ 0 \ \ 0], [0 \ \ 1 \ \ 0]\}$ in $\Re^3$.
(iii) $V = \{[1 \ \ 1], [1 \ \ 0]\}$ in $\Re^2$.

Solution

(i)   This is similar to the one discussed above except that the span of V will be the y axis in graphical notation.

Alternatively the spanning set is the set $\{y[0 \ \ 1]\}$ where y is any real number.

(ii)   If we take linear combinations of the two vectors given in $\Re^3$ we obtain vectors of the form,

$$x[1 \ \ 0 \ \ 0] + y[0 \ \ 1 \ \ 0]$$

Interpreting geometrically we are spanning the x–y plane in $\Re^3$.

(iii)   We have essentially given this example earlier. Linear combinations of the two vectors given will give all possible vectors in $\Re^2$ since we can get back to $\mathbf{j}$ by subtracting the two given vectors and thus we have $\mathbf{i}$ and $\mathbf{j}$, which span the whole of $\Re^2$, that is,

$$[1 \ \ 1] - [1 \ \ 0] = [0 \ \ 1]$$

Example 6B

Taking the three examples from Example 6A, consider whether the given set V, is a frame in the appropriate $\Re^n$. (Recall that Chapter 3 contains the definitions and ideas required.)

Solution

(i) $V = \{[0 \ \ 1]\}$ is a frame in $\Re^2$, since $\alpha[0 \ \ 1] = 0$ implies $\alpha = 0$.

(ii) $V = \{ [1 \ 0 \ 0], [0 \ 1 \ 0] \}$ is a frame in $\Re^3$, since

$$\alpha[1 \ \ 0 \ \ 0] + \beta[0 \ \ 1 \ \ 0] = \mathbf{0} = [0 \ \ 0 \ \ 0]$$

implies $\alpha = 0$ and $\beta = 0$. Alternatively, one vector is not a multiple of the other.

(iii) $V = \{[1 \ \ 1], [1 \ \ 0]\}$ is a frame in $\Re^2$, since one vector is not a multiple of the other.

Definition

A *basis* for a vector space V is a set of vectors which is both a frame **and** a span for V. This means that the set of vectors we have must be an independent set, and, further, when we take linear combinations of these independent vectors we generate the whole of the vector space V.

Example 6C

Consider again the three sets in the preceding two examples. Decide whether or not the set V is a basis for the appropriate vector space $\Re^n$.

Solution

(i) $V = \{[0 \ 1]\}$ is **not** a basis for $\Re^2$ since it is **not** a span for the whole of $\Re^2$. We note that V is a frame in $\Re^2$.

(ii) $V = \{[1 \ 0 \ 0], [0 \ 1 \ 0]\}$ is not a basis for $\Re^3$ since although it is a frame in $\Re^3$ it will not span the whole of $\Re^3$ but only the x–y plane contained in $\Re^3$.

(iii) $V = \{[1 \ 1], [1 \ 0]\}$ **is** a basis for $\Re^2$ since it is both a frame and a span, as seen in the preceding two examples.

6.5   THE DIMENSION OF A VECTOR SPACE

Definition

The number of vectors in a basis for a vector space is called *the dimension of the vector space*.

From this definition the reader will see that any basis for a vector space will have the same number of vectors in it.

From the preceding three examples we can see that $\Re^2$, which is a vector space, can have a basis given by the set $\{[1 \ 1], [1 \ 0]\}$.

Furthermore, we have seen that another basis for $\Re^2$ is given by the set $\{[1 \ 0], [0 \ 1]\}$. This second set consists of the two vectors **i** and **j** introduced earlier in the book:

This basis is known as *the standard basis* for $\Re^2$

We observe that although we have different bases they do contain the same number of vectors.

From the above we can now state that $\Re^2$ is a two-dimensional vector space over the scalars $\Re$.

In a similar fashion,

$\Re^n$ is an *n*-dimensional vector space over the scalars $\Re$

This means that we would need to find n independent vectors which span $\Re^n$ in order to provide a basis for $\Re^n$. If the basis of a vector space contains a finite number of vectors we refer to the vector space as being finite-dimensional. Otherwise, we have infinite-dimensional vector spaces.

Maximal Frame

From what we have seen in the above examples we can begin to think about what makes a frame a basis and what makes a span a basis. Starting with a frame in a certain vector space, we can put a further vector into our frame and examine whether or not this new set is still a frame or whether it has now become a dependent set. If it is still a frame we put another vector into the set and test again for independence or dependence. Sooner or later our new set will not be a frame if the vector space we are working in is finite-dimensional. At this point we have what is known as *a maximal frame*.

Minimal Span

Now let us start again, but this time with a span, V, for a certain vector space. We now ask the question, 'Is there a vector in V which is redundant?' This means that a linear combination of the remaining vectors can produce the redundant vector. If this is the case we throw out the so-called redundant vector. We continue in this way and see whether there is a further redundant vector. If there is we throw it out again. We can continue to do this until we come to the stage of having no more redundant vectors in our set. What we are left with is known as *a minimal span*.

Equivalent Definitions

Let V be a set of linearly independent vectors which span a certain vector space. Then the following three expressions are equivalent:

(i) V is a basis for the vector space.
(ii) V is a maximal frame in the vector space.
(iii) V is a minimal span in the vector space.

In practice it is often the case that we look for a maximal frame or a minimal span in order to obtain a basis for a vector space.

Example 6D

Why is the set given below **not** a basis for $\Re^2$?

$$V = \{[4 \ \ 0], [-1 \ \ 1], [3 \ \ -5]\}$$

Solution

The easiest answer to give would be that $\Re^2$ is a vector space with dimension 2. Therefore any basis for $\Re^2$ must contain exactly two vectors. As our set V has three vectors in it, it cannot be a basis.

Let us develop this question by using the ideas given above. The set V spans $\Re^2$ but is not a frame, since the equation given below can be solved uniquely, for α and β:

$$\alpha[4 \ \ 0] + \beta[-1 \ \ 1] = [3 \ \ -5]$$

This is equivalent to

$$[4\alpha \ \ 0] + [-\beta \ \ \beta] = [3 \ \ -5]$$

which in turns simplifies to

$$[4\alpha - \beta \ \ \beta] = [3 \ \ -5]$$

From this equation we see that

$$\beta = -5 \quad \text{and} \quad 4\alpha - \beta = 3$$

by equating coordinates. Hence,

$$\alpha = -\tfrac{1}{2} \quad \text{and} \quad \beta = -5$$

This means we have a true relationship of the form

$$-\tfrac{1}{2}[4 \ \ 0] - 5[-1 \ \ 1] = [3 \ \ -5]$$

At this stage we could say that the vector $[3 \ \ -5]$ is redundant since the remaining two vectors generate it.

However, we also observe that we could have made either of the first two vectors redundant because we could just simply rearrange the relationship into one of the following forms:

$$[4\ \ 0] = -10[-1\ \ 1] - 2[3\ \ -5]$$

or

$$[-1\ \ 1] = -\tfrac{1}{10}[4\ \ 0] - \tfrac{1}{5}[3\ \ -5]$$

Making the third vector redundant we now have the set

$$\{[4\ \ 0],\ [-1\ \ 1]\}$$

As one of these vectors is not a multiple of the other we conclude that one vector does not span the other. Therefore, we have a minimal span, that is, there are no more redundant vectors. Hence this set is a basis for $\Re^2$. Remember, bases are not unique.

Example 6E

Why is the set given below **not** a basis for $\Re^3$?

$$V = \{[1\ \ 0\ \ 0]\}$$

Solution

Again we could answer that $\Re^3$ is a vector space of dimension 3 and therefore any basis for $\Re^3$ must have three vectors in it. As the given set has only one vector in it, it cannot be a basis.

Let us again develop the ideas presented earlier in this chapter by using this example. The set V given above is a frame in $\Re^3$ and in fact spans a subspace of dimension 1. Therefore, according to our earlier discussion we can extend this set to obtain a maximal frame and thus a basis for $\Re^3$.

In this case the extension is particularly easy if we use the standard basis. We know that $[0\ \ 1\ \ 0]$ is independent of the given vector $[1\ \ 0\ \ 0]$ as it is not a multiple, so we can extend our set to

$$\{[1\ \ 0\ \ 0],\ [0\ \ 1\ \ 0]\}$$

This is still a frame. However, this set only spans the x–y plane in $\Re^3$ and therefore this set is still not a basis for $\Re^3$.

Now we seek a third vector which is independent of the two in this set. As we said above, the obvious choice is $[0\ \ 0\ \ 1]$. Thus we have the set

$$\{[1\ \ 0\ \ 0],\ [0\ \ 1\ \ 0],\ [0\ \ 0\ \ 1]\}$$

This is still a frame and it also has the right number of vectors so it must span $\Re^3$. It is the standard basis in $\Re^3$ and can be written as $\{i, j, k\}$, as introduced in Chapter 3.

Note that if we introduce a fourth vector into our set the frame turns into a dependent set and is thus no longer a maximal frame.

We now use DERIVE to answer similar questions to the above by recognizing the fact that some of the ideas introduced in earlier chapters can be used to our advantage.

DERIVE ACTIVITY 6A

(A)   Are the three vectors given below a basis for $\Re^3$?

$$[1 \ 1 \ 1], [-2 \ 1 \ 3], [5 \ 1 \ 0]$$

We note that we have the right number of vectors for a basis but we do not know whether the given vectors are independent. According to Chapter 3 we set up the equation

$$\alpha[1 \ 1 \ 1] + \beta[-2 \ 1 \ 3] + \gamma[5 \ 1 \ 0] = [0 \ 0 \ 0]$$

and try to show that the only solution is $\alpha = \beta = \gamma = 0$. If we cannot show this then the vectors would be dependent and a relationship could be found between them.

The vector equation above simplifies to

$$[\alpha \ \alpha \ \alpha] + [-2\beta \ \beta \ 3\beta] + [5\gamma \ \gamma \ 0] = [0 \ 0 \ 0]$$

This in turn reduces to the three equations

$$\alpha - 2\beta + 5\gamma = 0$$

$$\alpha + \beta + \gamma = 0$$

$$\alpha + 3\beta = 0$$

This is a system of three equations in three unknowns and, from Chapter 5, is an example of a homogeneous system of equations. In Chapter 5 we saw that such a system always has at least one solution, namely the trivial solution, $\alpha = \beta = \gamma = 0$. What we require to know is, 'Is this the only solution?'

In Chapter 5 we could answer this question by considering the rank of the matrix of coefficients, A, and the rank of the augmented matrix $(A|0)$.

Input the augmented matrix $(A|0)$ into DERIVE.

By using the ROW_REDUCE function find the rank of A and the augmented matrix. From this decide whether or not the solution for α, β and γ is the trivial one.

Your DERIVE screen should consist of just three lines as shown in Figure 6.1.

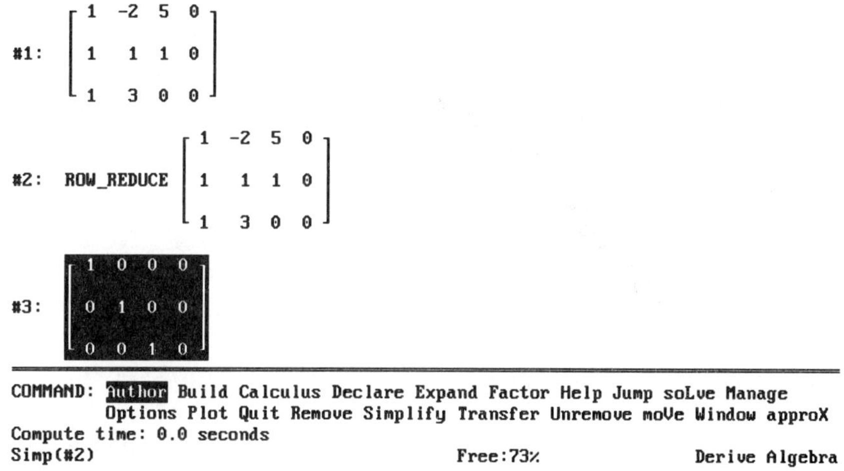

$$\#1: \quad \begin{bmatrix} 1 & -2 & 5 & 0 \\ 1 & 1 & 1 & 0 \\ 1 & 3 & 0 & 0 \end{bmatrix}$$

$$\#2: \quad \text{ROW\_REDUCE} \begin{bmatrix} 1 & -2 & 5 & 0 \\ 1 & 1 & 1 & 0 \\ 1 & 3 & 0 & 0 \end{bmatrix}$$

$$\#3: \quad \begin{bmatrix} 1 & 0 & 0 & 0 \\ 0 & 1 & 0 & 0 \\ 0 & 0 & 1 & 0 \end{bmatrix}$$

COMMAND: Author Build Calculus Declare Expand Factor Help Jump soLve Manage
              Options Plot Quit Remove Simplify Transfer Unremove moVe Window approX
Compute time: 0.0 seconds
Simp(#2)   Free:73%   Derive Algebra

Figure 6.1   The rank of the matrix is 3

From the fact that the rank of A is 3 and $n = 3$, Chapter 5 tells us that the solution to our three equations is indeed unique and therefore,

$$\alpha = \beta = \gamma = 0$$

Thus the three given vectors are a maximal frame and thus a basis for $\Re^3$.

We can cut the working down even further if we recall that the column rank is equal to the row rank of a matrix.

Observe that the first three columns of our augmented matrix are simply the three vectors we were originally given.

Hence we could just take the three vectors and put them into the first three columns of a matrix with the fourth column full of zeros. In this case we do not need to even consider α, β and γ.

(B)   Are the three vectors given below a basis for $\Re^3$?

$$[2 \ 1 \ 0], [8 \ -10 \ -6], [0 \ 7 \ 3]$$

Declare a 3×4 matrix again and input directly into DERIVE the three vectors with the zero vector as columns. Again work out the rank and make your conclusion. The DERIVE screen is shown in Figure 6.2.

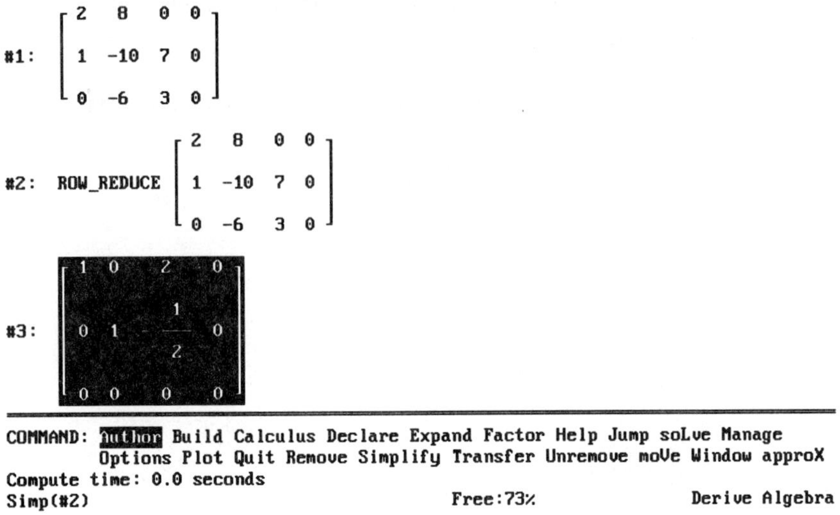

#1: $\begin{bmatrix} 2 & 8 & 0 & 0 \\ 1 & -10 & 7 & 0 \\ 0 & -6 & 3 & 0 \end{bmatrix}$

#2: ROW_REDUCE $\begin{bmatrix} 2 & 8 & 0 & 0 \\ 1 & -10 & 7 & 0 \\ 0 & -6 & 3 & 0 \end{bmatrix}$

#3: $\begin{bmatrix} 1 & 0 & 2 & 0 \\ 0 & 1 & -\dfrac{1}{2} & 0 \\ 0 & 0 & 0 & 0 \end{bmatrix}$

COMMAND: Author Build Calculus Declare Expand Factor Help Jump soLve Manage
         Options Plot Quit Remove Simplify Transfer Unremove moVe Window approX
Compute time: 0.0 seconds
Simp(#2)   Free:73%   Derive Algebra

Figure 6.2   The rank of the matrix is 2

Interpreting line 3 we see that the rank of A is 2 and therefore the rank of the augmented matrix is also 2, as stated in Chapter 5. The conclusion is that there is an infinity of solutions and therefore if we had introduced α, β and γ as before, we would say that they are **not** uniquely zero. This means that the given vectors are dependent and therefore cannot be a basis for $\Re^3$.

At this stage we would like to write down a relationship between the three vectors, for such a relationship must now exist. The relationship we have assumed in the above work is

$$\alpha[2\ 1\ 0] + \beta[8\ -10\ -6] + \gamma[0\ 7\ 3] = [0\ 0\ 0]$$

Interpreting line 3 again with α, β and γ in mind, we have from the first row of the matrix the equation

$$\alpha + 2\gamma = 0$$

and from the second row

$$\beta - \tfrac{1}{2}\gamma = 0$$

Hence α, and β can be written in terms of γ. Thus,

$$\alpha = -2\gamma, \quad \beta = \tfrac{1}{2}\gamma$$

Putting these back into the relationship equation we have

$$-2\gamma[2 \ \ 1 \ \ 0] + \tfrac{1}{2}\gamma[8 \ \ -10 \ \ -6] + \gamma[0 \ \ 7 \ \ 3] = [0 \ \ 0 \ \ 0]$$

As γ can be any value we choose it to be 2 in order to rid the equation of fractions. Thus we have

$$-4[2 \ \ 1 \ \ 0] + [8 \ \ -10 \ \ -6] + 2[0 \ \ 7 \ \ 3] = [0 \ \ 0 \ \ 0]$$

Alternatively, by rearranging we have

$$[8 \ \ -10 \ \ -6] = 4[2 \ \ 1 \ \ 0] - 2[0 \ \ 7 \ \ 3]$$

So what we have shown is that three lines in DERIVE can give us all the information we need to know whether sets of vectors are dependent or independent. Furthermore, by correct interpretation of the simplified reduced matrix we can write down in a few lines the relationship between the vectors (if they are dependent on each other).

Column and Row Spaces of a Matrix

Definition

The *column space* of a matrix A is the set of all linear combinations of the columns of A.

This means that we are looking for the space spanned or generated by the columns of a matrix considered as vectors. This is exactly what we were doing in DERIVE Activity 6A. In (A) the column space was $\Re^3$, whereas in (B) the column space was two-dimensional and thus a subspace of $\Re^3$.

In a similar way we can define the row space of a matrix A.

Definition

The *row space* of a matrix A is the set of all linear combinations of the rows of A.

Because of rank considerations, the dimension of the column space of a particular matrix is equal to the dimension of the row space of that matrix.

DERIVE ACTIVITY 6B

(A)   Determine whether the matrices given below have the same row space:

$$A = \begin{bmatrix} 1 & 1 & 5 \\ 2 & 3 & 13 \end{bmatrix} \qquad B = \begin{bmatrix} 1 & -1 & -2 \\ 3 & -2 & -3 \end{bmatrix} \qquad C = \begin{bmatrix} 1 & -1 & -1 \\ 4 & -3 & -1 \\ 3 & -1 & 3 \end{bmatrix}$$

Input each of the matrices into DERIVE.   Row-reduce each of them and compare the results.

You should find that A and C reduce to the same row vectors whereas B does not.

Explicitly you should obtain

$$\text{Reduced } A = \begin{bmatrix} 1 & 0 & 2 \\ 0 & 1 & 3 \end{bmatrix}$$

$$\text{Reduced } B = \begin{bmatrix} 1 & 0 & 1 \\ 0 & 1 & 3 \end{bmatrix}$$

and

$$\text{Reduced } C = \begin{bmatrix} 1 & 0 & 2 \\ 0 & 1 & 3 \\ 0 & 0 & 0 \end{bmatrix}$$

From these results you can interpret that the row space of A is two-dimensional and is spanned by the two vectors [1 0 2] and [0 1 3], whereas the row space of B, although it is also two-dimensional, is spanned by [1 0 1] and [0 1 3].

Geometrically the row space of A and the row space of B are two planes which intersect in a common line, spanned by the vector [0 1 3].

Finally, the row space of C is two-dimensional because the third vector (or row), [3 −1 3], was eliminated by the first two rows to obtain the row of zeros displayed.   Furthermore, the two basis vectors for its row space are identical to those of the row space of matrix A, and hence our conclusion.

(B)   Determine whether or not the following matrices have the same column space:

$$A = \begin{bmatrix} 1 & 3 & 5 \\ 1 & 4 & 3 \\ 1 & 1 & 9 \end{bmatrix} \qquad B = \begin{bmatrix} 1 & 2 & 3 \\ -2 & -3 & -4 \\ 7 & 12 & 17 \end{bmatrix}$$

Since we are asked to consider the column space and we do not just want the dimension of the spaces, we must transpose the matrices before we use DERIVE, or input them and then transpose, since DERIVE only row-reduces.

Note that if we simply worked out the dimension of the respective column spaces and they turned out to be different, we would have finished the question, as the column spaces could not be the same on dimension grounds alone. If, however, we find that the column ranks are the same we would still have to transpose and row-reduce to find whether the bases are the same. With DERIVE it is just as simple to do the transposing and reduction from the start.

In the case we are considering, you should find that A' and B', the transposes of A and B, reduce to the same matrix, namely,

$$\begin{bmatrix} 1 & 0 & 3 \\ 0 & 1 & -2 \\ 0 & 0 & 0 \end{bmatrix}$$

Hence we can conclude that the column spaces are the same.

The common column space is a two-dimensional subspace of $\Re^3$ and a basis for it could be

$$[1 \ 0 \ 3], [0 \ 1 \ -2]$$

It should be clear to you that we have given more than a straight answer to the original problem.

We hope that, by so doing, you have a better insight into what is happening, and also appreciate the power that DERIVE can bring to bear on such problems.

6.6   THE NULL SPACE OF A MATRIX

In this section we introduce a new word but not a new idea. Recall again a system of homogeneous equations, as we had, for example, in DERIVE Activity 6A above and as introduced in Chapter 5, Section 5.3.

Such a system can be written

$$A\mathbf{x} = \mathbf{0}$$

Further, we observed in Chapter 5 that this system of equations always has a solution, namely, the trivial one, $\mathbf{x} = \mathbf{0}$.

Definition

The *null space* of a matrix A is the set of solutions to the equation $A\mathbf{x} = \mathbf{0}$. Our first observation is that this space is non-empty since it must contain the zero vector, $\mathbf{0}$. We now prove that the null space of a matrix is in fact a vector subspace of the vector space $\Re^n$ for suitable n. The proof uses the three criteria for a subspace given in Section 6.3 above.

Fact

The null space of an $n \times n$ matrix A is a subspace of $\Re^n$.

Proof

First we know that the null space is not empty because it contains the trivial solution $\mathbf{x} = \mathbf{0}$. Thus the first criterion is satisfied.

Finally, we need to prove that the null space is closed under addition and scalar multiplication. This means that we have to show that if $\mathbf{x}_1$ and $\mathbf{x}_2$ are solutions of $A\mathbf{x} = \mathbf{0}$, that is, $A\mathbf{x}_1 = \mathbf{0}$ and $A\mathbf{x}_2 = \mathbf{0}$, then $\mathbf{x}_1 + \mathbf{x}_2$ and $\alpha\mathbf{x}_1$ are also solutions of the equation $A\mathbf{x} = \mathbf{0}$, that is, $A(\mathbf{x}_1 + \mathbf{x}_2) = \mathbf{0}$ and $A(\alpha\mathbf{x}_1) = \mathbf{0}$.

We prove these statements by appealing to the properties of matrices introduced earlier in the book. Consider first the expression

$$A(\mathbf{x}_1 + \mathbf{x}_2)$$

This expands to give

$$A(\mathbf{x}_1 + \mathbf{x}_2) = A\mathbf{x}_1 + A\mathbf{x}_2$$

But the right-hand side is simply $\mathbf{0}$ since $A\mathbf{x}_1 = \mathbf{0}$ and $A\mathbf{x}_2 = \mathbf{0}$. Hence we have shown that

$$A(\mathbf{x}_1 + \mathbf{x}_2) = \mathbf{0}$$

Similarly, consider the expression $A(\alpha\mathbf{x}_1)$. Expanding:

$$A(\alpha\mathbf{x}_1) = \alpha\,A(\mathbf{x}_1)$$

But again since $A(\mathbf{x}_1) = 0$ the right-hand side will be 0 and thus we have shown that

$$A(\alpha\mathbf{x}_1) = 0$$

Hence we have shown that the null space of a matrix is a vector space in its own right and is a subspace of the vector space $\mathfrak{R}^n$ since all the $\mathbf{x}_i$ are $n \times 1$ column vectors.

We can generalize this idea to a non-square matrix.

If A is an $m \times n$ matrix and $\mathbf{x}$ is an $n \times 1$ matrix then $A\mathbf{x}$ is an $m \times 1$ matrix. The null space of the matrix A will be a subspace of the vector space $\mathfrak{R}^n$.

Example 6F

Is the vector $[1 \quad 3 \quad -4]$ in the null space of the matrix A given below?

$$A = \begin{bmatrix} 3 & -5 & -3 \\ 6 & -2 & 0 \\ -8 & 4 & 1 \end{bmatrix}$$

Solution

This type of question is direct and simply needs us to test whether or not $A\mathbf{x} = 0$.

Using DERIVE or by hand compute $A\mathbf{x}$:

$$A \cdot \mathbf{x} = \begin{bmatrix} 3 & -5 & -3 \\ 6 & -2 & 0 \\ -8 & 4 & 1 \end{bmatrix} \cdot \begin{bmatrix} 1 \\ 3 \\ -4 \end{bmatrix}$$

You should find that this gives the zero vector, 0, and hence we can conclude that the given vector is in the null space of A.

Example 6G

Give an explicit description of the null space of the following matrix:

$$A = \begin{bmatrix} 1 & -2 & 0 & 4 & 0 \\ 0 & 0 & 1 & -9 & 0 \\ 0 & 0 & 0 & 0 & 1 \end{bmatrix}$$

Solution

This is a much more difficult question than that in Example 6F since we have to find the null space by solving the equation $A\mathbf{x} = \mathbf{0}$.

By matrix multiplication the vectors $\mathbf{x}$ must be in $\Re^5$, since A is 3×5 and thus $\mathbf{x}$ must be 5×1.

In this exercise you do not need to use DERIVE, as the matrix A is already in reduced echelon form and is thus ready to interpret.

Let $\mathbf{x} = [x_1 \ x_2 \ x_3 \ x_4 \ x_5]$. Write down explicitly the equations in x_i obtained from

$$A\mathbf{x} = \mathbf{0}$$

From the third equation we obtain

$$x_5 = 0$$

From the second equation we have

$$x_3 - 9x_4 = 0$$

Finally, from the first equation we have

$$x_1 - 2x_2 + 4x_4 = 0$$

From these three equations we conclude that the fifth coordinate of $\mathbf{x}$ must always be zero. Further,

$$x_3 = 9x_4 \quad \text{and} \quad x_1 = 2x_2 - 4x_4$$

Thus any solution to $A\mathbf{x} = \mathbf{0}$ must be of the form

$$[2x_2 - 4x_4 \; x_2 \; 9x_4 \; x_4 \; 0]$$

This is known as the general solution as it gives **all** the possible solutions to the equation.

Notice that there are two variables on which the first four coordinates rely. In the literature these are known as *free variables* or *independent variables* and the number of them is often referred to as the *number of degrees of freedom* of the system of equations. Thus in this exercise we have two free variables, or a system of equations with two degrees of freedom. In simple terms this means that we have two free variables to each of which we can assign a value and thus obtain a solution to our equation. We have two degrees of freedom, that is, two choices.

Take $x_2 = 0$ and $x_4 = 1$, for example, and we would have the solution

$$[-4 \; 0 \; 9 \; 1 \; 0]$$

Choose again! Take $x_2 = 1$ and $x_4 = 0$ for example. Then the solution to our equation is

$$[2 \; 1 \; 0 \; 0 \; 0]$$

Clearly, there will be an infinity of solutions to our problem as we can choose x_2 and x_4 to be any values. Furthermore, the solutions form a vector space, the null space.

The variables which are not free are often referred to as the *leading variables* or the *dependent variables* or the *basic variables*. We note that the leading variables have coefficients of 1 because of the row-reduced echelon form of the matrix, and they are the pivot elements introduced in Chapter 5.

In summary, so far we have leading variables x_1, x_3 and x_5, and free variables x_2 and x_4. Furthermore, the general solution vector

$$[x_1, x_2, x_3, x_4, x_5]$$

can be written as

$$[x_1, x_2, x_3, x_4, x_5] = [2x_2 - 4x_4, x_2, 9x_4, x_4, 0]$$
$$= x_2[2, 1, 0, 0, 0] + x_4[-4, 0, 9, 1, 0]$$

This shows that every vector **x** which satisfies $A\mathbf{x} = \mathbf{0}$ is generated (or spanned) by a linear combination of the vectors:

$$[2, 1, 0, 0, 0] \quad \text{and} \quad [-4, 0, 9, 1, 0]$$

and these are the two we generated by substituting different values for x_2 and x_4 above.

As these two vectors are independent they form a basis for the null space and thus the dimension of the null space is two.

From what we have done so far in this exercise we can state the following facts without further proof.

Three Facts

(i)   The number of free variables in the equation $A\mathbf{x} = \mathbf{0}$ is the dimension of the null space of A.

(ii)   The spanning vectors we obtain by this method are automatically independent since in our case, the linear combination

$$x_2[2, 1, 0, 0, 0] + x_4[-4, 0, 9, 1, 0] = [2x_2 - 4x_4, x_2, 9x_4, x_4, 0]$$
$$= [0, 0, 0, 0, 0]$$

implies x_2 and x_4 are zero.

This will always be the case as the free variables will be single coordinates in the solution vector, in our case the second and fourth coordinates. Thus when we apply the criterion for independence and equate the linear combination to the zero vector those free variables will automatically be zero.

(iii)   In general, if n is the number of variables in a system of homogeneous equations and r is the rank of the coefficient matrix, then there will be r leading variables and $(n - r)$ free variables. Then the dimension of the null space will also be $(n - r)$, from fact (i) above.

Exercise 6C

The following matrix M represents a linear mapping from $\mathfrak{R}^3$ to $\mathfrak{R}^4$:

$$M = \begin{bmatrix} 1 & 2 & 1 \\ 2 & 5 & 3 \\ 5 & 14 & 9 \\ 7 & 18 & 11 \end{bmatrix}$$

(i)   Is the vector $[1 \ 0 \ 1]$ in the null space of the matrix M?
(ii)   Find the null space of the matrix A.

We finish this section with a DERIVE activity to reinforce these important ideas.

DERIVE ACTIVITY 6C

(A)   For the matrix A, given below, work out the dimension of its null space and find a basis for it:

$$A = \begin{bmatrix} 1 & 2 & -5 & 11 & -3 \\ 2 & 4 & -5 & 15 & 2 \\ 1 & 2 & 0 & 4 & 5 \\ 3 & 6 & -5 & 19 & -2 \end{bmatrix}$$

Input the matrix into DERIVE and row-reduce the matrix A. (Remember, DERIVE reduces matrices to 'row-reduced echelon form' and that is the form we must have before we can interpret the results.)

The third line of your session in DERIVE should have produced the following reduced matrix:

$$\begin{bmatrix} 1 & 2 & 0 & 4 & 0 \\ 0 & 0 & 1 & -7/5 & 0 \\ 0 & 0 & 0 & 0 & 1 \\ 0 & 0 & 0 & 0 & 0 \end{bmatrix}$$

We have in mind the system of equations $A\mathbf{x} = \mathbf{0}$ in which we have four equations in five unknown variables, denoted by x_1, x_2, x_3, x_4, and x_5; hence $n = 5$.

From the reduced matrix we can see that the rank, r, is 3. Hence, $(n - r)$ is 2 and thus we have two free variables. Therefore, the dimension of the null space is also 2.

The pivots or leading variables are x_1, x_3 and x_5 and hence the free variables are x_2 and x_4. Finally, we interpret the reduced equations and obtain

$$x_1 + 2x_2 + 4x_4 = 0$$

$$x_3 - \tfrac{7}{5}x_4 = 0$$

$$x_5 = 0$$

Hence the general solution is

$$[x_1, x_2, x_3, x_4, x_5] = [-2x_2 - 4x_4, x_2, \tfrac{7}{5}x_4, x_4, 0]$$

$$= x_2[-2, 1, 0, 0, 0] + x_4[-4, 0, \tfrac{7}{5}, 1, 0]$$

Thus a basis for the null space could be

$$\{[-2, 1, 0, 0, 0], [-4, 0, \tfrac{7}{5}, 1, 0]\}$$

Note that we could have generated these basis vectors by choosing the free variables in the general solution as $x_2 = 1$ and $x_4 = 0$ and, secondly, as $x_2 = 0$ and $x_4 = 1$.

(B)   Repeat activity (A) for the following matrices:

$$\begin{bmatrix} 1 & 2 & 1 \\ 2 & 5 & 3 \\ 5 & 14 & 9 \\ 7 & 18 & 11 \end{bmatrix} \qquad \begin{bmatrix} 1 & 2 & -3 & 0 & 4 & 6 \\ -1 & 15 & 4 & 1 & 2 & -1 \\ 3 & 6 & 2 & -1 & -3 & 1 \\ 1 & 0 & 5 & 2 & -1 & 4 \\ 2 & 7 & -11 & 0 & 1 & 2 \end{bmatrix}$$

6.7   LINEAR TRANSFORMATIONS

We begin this section by introducing the idea of a function, although we hope that many readers will already have met such a concept.

Definition

Let X and Y be two sets and let f be a rule that takes an element of X and associates it with a unique element of Y. Then we will call f a *function*.

Notation

We speak of a function f from X to Y and write $f\colon X \to Y$. The arrow $\to$ is used between the two sets X and Y. If $x \in X$ and the associated element in Y is y we write $x \mapsto y$. The arrow $\mapsto$ is used between elements. We also write $f(x) = y$ as the functional relationship. Combining all these we have the notation

$$f: X \to Y$$

$$x \mapsto y = f(x)$$

Definitions

The set X is called the *domain of the function.*
The set Y is called the *codomain of the function.*
The set $f(X) = \{f(x): x \in X\}$ is called the *image (or range) of the function.*
    We observe that the image is a subset of the codomain, that is, $f(X) \subseteq Y$.
    Examples of familiar functions are:

1.   The function $f: \Re \to \Re$ is defined by the relation $f(x) = x^2$. The domain and codomain are the set of real numbers. The function f takes a real number and associates with it its square, which is unique. Notice that there are some elements in the codomain which will never be associated with an element in the domain, for example the negative numbers. Therefore, the image of f is the set of positive real numbers together with zero. This set is a subset of the codomain $\Re$.

2.   The function, $f: \Re \to [-1, +1]$ is defined by the relation

$$f(x) = \sin(x)$$

We could have simply written the function as follows:

$$\sin: \Re \to [-1, +1]$$

This gives us all the following information:

    (i)   The domain of 'sin' is the set of real numbers.
    (ii)   the codomain of 'sin' is the closed interval from -1 to $+1$.
    (iii)   the images are $\sin(x)$ for $x \in \Re$.
    (iv)   by drawing the graph of this function, or otherwise, we can conclude that the image and the codomain are equal.

3.   The function; DET: $M_n \to \Re$ defined by the relation

'DET(A) is the determinant of the $n \times n$ square matrix A'

Here we have used M_n to stand for the $n \times n$ square matrices, and, as we have seen in Chapter 2, we can take determinants of such matrices. We say that the function is well defined. Furthermore, the value of a determinant is a real number.

Note that we could let a determinant have entries from the complex numbers, in which case we would redefine our function with a different codomain.

Note also in this example that the image of this function is impossible to state. All we know is that for a given square matrix we can work out its determinant and say that it is a real number. As our codomain contains all the real numbers we know that we have defined the determinant function well.

In summary, the codomain is a large set in which we know that all our images will lie. It is only in certain functions that we can actually say what the image set will be.

Now let us consider a function which has for its domain a vector space V and for its codomain another vector space W. In a special case V could equal W. We will label the function T and refer to it in what follows as a transformation, so

$$T: V \rightarrow W$$

Now since V is a vector space it has two operations associated with it, namely, vector addition and scalar multiplication. (If you need to revise these ideas refer to Section 6.1.)

Consider the sum of two vectors, $v_1 + v_2$ where $v_1 \in V$ and $v_2 \in V$. Recall that by closure $v_1 + v_2 \in V$. Also $\alpha v_1 \in V$ for any scalar α. Therefore, $T(v_1)$, $T(v_2)$, $T(v_1 + v_2)$ and $T(\alpha v_1)$ are all well defined and belong to the vector space W. Let us put $T(v_1) = w_1$; $T(v_2) = w_2$; and $T(v_1 + v_2) = w_3$ to indicate the above fact.

Symbolically we write

$$T: \quad V \rightarrow W$$

$$v_1 \, \alpha \, T(v_1) = w_1$$

$$v_2 \, \alpha \, T(v_2) = w_2$$

$$v_1 + v_2 \, \alpha \, T(v_1 + v_2) = w_3$$

$$\alpha v_1 \, \alpha \, T(\alpha v_1)$$

As always in mathematics we should start asking questions, seeking patterns and suggesting, 'wouldn't it be nice if ...'. In this case since v_1 and v_2 are added to make $v_1 + v_2$ wouldn't it be nice if their associated images, which are vectors in W, behaved the same? That would mean

$$T(v_1) + T(v_2) = T(v_1 + v_2)$$

or

$$\mathbf{w}_1 + \mathbf{w}_2 = \mathbf{w}_3$$

Also, since we can multiply the vector $\mathbf{v}_1$ by α to obtain $\alpha\mathbf{v}_1$ wouldn't it be nice if

$$T(\alpha\mathbf{v}_1) = \alpha T(\mathbf{v}_1) = \alpha\mathbf{w}_1?$$

These two ideas combine to give us the definition of a linear transformation.

Definition

If $T: V \to W$, where V and W are vector spaces, satisfies the following two axioms, we say that T is a *linear transformation*:

(i) $T(\mathbf{v}_1 + \mathbf{v}_2) = T(\mathbf{v}_1) + T(\mathbf{v}_2)$ for all $\mathbf{v}_1$ and $\mathbf{v}_2 \in V$.

(ii) $T(\alpha\mathbf{v}_1) = \alpha T(\mathbf{v}_1)$ for all $\mathbf{v}_1 \in V$ and α any scalar.

Alternative Definition

Instead of the two axioms above we can replace them with the single axiom

$$T(\alpha_1\mathbf{v}_1 + \alpha_2\mathbf{v}_2) = \alpha_1 T(\mathbf{v}_1) + \alpha_2 T(\mathbf{v}_2)$$

where $\mathbf{v}_1$ and $\mathbf{v}_2$ belong to V and α_1 and α_2 are any two scalars. The proof that these two definitions are equivalent is given below.

Assume that the alternative definition holds; then choosing α_1 and α_2 as the unit scalar, axiom (i) follows. Also, choosing α_2 equal to the zero scalar, axiom (ii) follows.

Now assume that the two axioms hold, and consider

$$T(\alpha_1\mathbf{v}_1 + \alpha_2\mathbf{v}_2)$$

By using axioms (i) and (ii), respectively, we have

$$T(\alpha_1\mathbf{v}_1 + \alpha_2\mathbf{v}_2) = T(\alpha_1\mathbf{v}_1) + T(\alpha_2\mathbf{v}_2) = \alpha_1 T(\mathbf{v}_1) + \alpha_2 T(\mathbf{v}_2)$$

This proves that the two definitions are equivalent.

The first axiom says that vector addition is preserved, while the second axiom says that scalar multiplication is preserved. In the alternative definition we would say that linear combinations are preserved. In either case we talk of structure preserving transformations. This idea is found in other branches of mathematics, for example group theory.

We now show how linear transformations and matrices can be interchangeable. This means that a matrix can be interpreted as a linear transformation, and that a linear transformation between finite-dimensional vector spaces can be written as a matrix. We have already used the matrix equation $A\mathbf{x} = \mathbf{k}$ several times in this book, where A is an $m \times n$ matrix with $\mathbf{x}$ being an $n \times 1$ matrix and $\mathbf{k}$ an $m \times 1$ matrix, both $\mathbf{x}$ and $\mathbf{k}$ being considered as column vectors. Furthermore, we have seen that $\mathfrak{R}^n$ for any positive integer n can be considered as a vector space with appropriate definitions of vector addition and scalar multiplication.

Let us consider a function $f: \mathfrak{R}^n \to \mathfrak{R}^m$, where an element of the domain can be considered as a column vector with n coordinates. The image $f(\mathbf{x})$ will be an element of $\mathfrak{R}^m$ and can therefore be considered as a column vector with m coordinates.

Define f by the relationship, $f(\mathbf{x}) = A\mathbf{x}\ (=\mathbf{k})$, where A is an $m \times n$ matrix.

Note that A has to be $m \times n$ and **not** $n \times m$ for conformability with $\mathbf{x}$. Also note how n and m relate to the dimensions of the domain and codomain.

The fact that f with this definition is a linear transformation follows from the alternative definition, together with matrix properties from Chapter 1, as indicated below. We have to show that

$$f(\alpha_1 \mathbf{x}_1 + \alpha_2 \mathbf{x}_2) = \alpha_1 f(\mathbf{x}_1) + \alpha_2 f(\mathbf{x}_2)$$

Consider, then,

$$f(\alpha_1 \mathbf{x}_1 + \alpha_2 \mathbf{x}_2) = A(\alpha_1 \mathbf{x}_1 + \alpha_2 \mathbf{x}_2)$$

by the definition above.

Now,

$$A(\alpha_1 \mathbf{x}_1 + \alpha_2 \mathbf{x}_2) = A(\alpha_1 \mathbf{x}_1) + A(\alpha_2 \mathbf{x}_2)$$

by a matrix property and

$$A(\alpha_1 \mathbf{x}_1) + A(\alpha_2 \mathbf{x}_2) = \alpha_1 A(\mathbf{x}_1) + \alpha_2 A(\mathbf{x}_2)$$

by another property.

Finally,

$$\alpha_1 A(\mathbf{x}_1) + \alpha_2 A(\mathbf{x}_2) = \alpha_1 f(\mathbf{x}_1) + \alpha_2 f(\mathbf{x}_2)$$

by the definition of f.

Hence we have shown that f is a linear transformation, as claimed.

We will now examine linear transformations and see whether or not some of the concepts introduced in this chapter for matrices transfer over. For example, in matrices we had the idea of a null space. What, if anything, does this look like in linear transformations? We answer this and other such questions in the next section.

Before we do that we give some examples of linear transformations and some exercises to consolidate these ideas.

Exercise 6D

Examples of linear transformations

1. The identity linear transformation.
    Let $T: V \rightarrow V$ be defined by $T(\mathbf{v}) = \mathbf{v}$, for every $\mathbf{v}$ in V.
    Check the two axioms or the alternative axiom to show that this is indeed a linear transformation.
    What would this transformation look like if written in matrices?
    Would we have to impose conditions on our original transformation?
2. The zero linear transformation.
    Let $T: V \rightarrow W$ be defined by $T(\mathbf{v}) = \mathbf{0}$, for every $\mathbf{v}$ in V, where $\mathbf{0}$ is the zero vector in W.
    Show that this is a linear transformation.
    What would this transformation look like if written in matrices?
    Would we have to impose conditions on our original transformation?
3. Geometrical transformations.
    (i)   Reflection in the x-axis.
           Let $T: \Re^2 \rightarrow \Re^2$ be defined by $T([x, y]) = [x, -y]$.
           Prove that this is a linear transformation.
           This transformation can be seen graphically to take a point in $\Re^2$ (the plane) and associate with it the point reflected in the x-axis. The appropriate matrix would be 2×2 because of the domain and codomain.
           Explicitly we would have

           $$\begin{bmatrix} 1 & 0 \\ 0 & -1 \end{bmatrix}$$

           If we call this matrix A, check that

           $$A \cdot \begin{bmatrix} x \\ y \end{bmatrix} = \begin{bmatrix} x \\ -y \end{bmatrix}$$

(ii)   Reflection in the *y*-axis.

Let $T: \Re^2 \to \Re^2$ be defined by $T([x, y]) = [-x, y]$.

Follow through similar steps to those given in (i) above to provide an explicit matrix which has the same effect as the linear transformation.

(iii)   Reflection in the line $y = x$.

Let $T: \Re^2 \to \Re^2$ be defined by $T([x, y]) = [y, x]$.

Again follow the same procedure to satisfy yourself that this is a linear transformation, and find a matrix which has the same effect.

(iv)   Rotation about the origin anti-clockwise through 90°.

Let $T: \Re^2 \to \Re^2$ be defined by $T([x, y]) = [-y, x]$.

Again follow the same procedure to satisfy yourself that this is a linear transformation, and find a matrix which has the same effect.

(v)   Define linear transformations for yourself which produce the following results:

(a)   Rotation about the origin through 180°.

(b)   Rotation about the origin clockwise through 90°.

(c)   Reflection in the line $y = -x$.

(vi)   In this exercise we will generalize rotations about the origin.

From the geometry of the situation write down a 2 × 2 matrix which transforms a vector $[x, y]$ to another vector in $\Re^2$ which is obtained from $[x, y]$ by a rotation anti-clockwise about the origin through $\theta°$.

You are strongly encouraged to do this for yourself and then compare your result with the matrix given below.

For linear transformations from $\Re^2$ to $\Re^2$, there is a short cut for writing down the appropriate 2 × 2 matrix. The idea is to choose the standard basis for $\Re^2$, namely,

$$\{[1, 0], [0, 1]\}$$

These are simply the vectors **i** and **j** introduced in Chapter 3. If we follow where these vectors go to under a certain transformation we can immediately write down the appropriate matrix for the transformation.

Assume that the matrix is

$$\begin{bmatrix} a & b \\ c & d \end{bmatrix}$$

Then $[1, 0]$ has $[a, c]$ for its image and $[0, 1]$ has $[b, c]$ for its image. Hence, knowing where the basis vectors go to we can write down the matrix as claimed. Show that this result works for parts (i)–(v).

Now we give a solution to part (iv).

Solution for the General Rotation

If you draw the x and y axes on a piece of paper and then assume that the point $[1, 0]$ moves $\theta°$ round the origin anti-clockwise to a new position, you should be able to work out the coordinates for the image. Remember that the length from the origin to this image point is still 1 because we are moving on the unit circle. You should find that

$$[1, 0] \mapsto [\cos(\theta), \sin(\theta)]$$

Similarly, you should find that

$$[0, 1] \mapsto [-\sin(\theta), \cos(\theta)]$$

The resulting general matrix is therefore

$$\begin{bmatrix} \cos(\theta) & -\sin(\theta) \\ \sin(\theta) & \cos(\theta) \end{bmatrix}$$

We finish this section with examples of transformations which are not linear.

(i)   Let $T: \mathfrak{R}^2 \to \mathfrak{R}^2$ be defined by $T([x, y]) = [x + 1, y]$.
       We might think this looks linear but if we test the first axiom we obtain

$$T([x_1, y_1] + [x_2, y_2]) = T([x_1 + x_2, y_1 + y_2])$$

$$= [x_1 + x_2 + 1, y_1 + y_2]$$

whereas

$$T([x_1, y_1]) + T([x_2, y_2]) = [x_1 + 1, y_1] + [x_2 + 1, y_2]$$

$$= [x_1 + x_2 + 2, y_1 + y_2]$$

Hence,

$$T([x_1, y_1] + [x_2, y_2]) \neq T([x_1, y_1]) + T([x_2, y_2])$$

An alternative, but less elegant, approach would be to try to find a matrix which sends $[x, y]$ to $[x + 1, y]$, that is, find A such that

$$A \cdot [x, y]^T = [x + 1, y]^T$$

Of course, from what we have just shown there is no 2×2 matrix with real number entries which satisfies the equation, so we could never produce one. The problem is the addition of 1 to the x coordinate.

　　　　　The second transformation which is not linear, given below, is more obvious.

(ii)　　Let $T: \mathfrak{R}^2 \to \mathfrak{R}^2$ be defined by $T([x, y]) = [x^2, y]$.

　　　　　Our first reaction might be that a coordinate which is a square number could not preserve linearity. Our hunch would be correct.

　　　　　Again the first axiom fails since

$$T([x_1, y_1] + [x_2, y_2]) = T([x_1 + x_2, y_1 + y_2])$$

$$= [(x_1 + x_2)^2, y_1 + y_2]$$

whereas

$$T([x_1, y_1]) + T([x_2, y_2]) = [x_1^2, y_1] + [x_2^2, y_2]$$

$$= [x_1^2 + x_2^2, y_1 + y_2]$$

Hence since $(x_1 + x_2)^2 \neq x_1^2 + x_2^2$ the first axiom for a linear transformation doesn't hold, as claimed.

In the next section we fulfil our promise to look at parallels between linear transformations and matrices.

We begin with the counterpart of the null space.

6.8　KERNEL AND IMAGE OF A LINEAR TRANSFORMATION

We observe from the first axiom for a linear transformation that the zero vector in the domain always has the zero vector of the codomain for its image. This means that, given a linear transformation, $T: V \to W$, then

$$T(0_v) = 0_w$$

where 0_v and 0_w are the respective zero or null vectors for the vector spaces V and W. This statement is verified by observing that

$$T(0_v) = T(0_v + 0_v) = T(0_v) + T(0_v)$$

by the first axiom. Note that this is an equation concerning vectors in W.

Now subtract $T(0_v)$ from both sides of the equation obtained above, namely,

$$T(0_v) = T(0_v) + T(0_v)$$

and we obtain

$$0_w = T(0_v)$$

as claimed.

There may be other vectors in the domain of a linear transformation which have the zero vector of the codomain as their image.

Below we define the set of all vectors in the domain that have the zero vector in the codomain as image.

Definition

The *kernel of a linear transformation*, $T: V \rightarrow W$, written Ker(T), is defined by

$$\text{Ker}(T) = \{v \in V : T(v) = 0_w\}$$

Since we defined $T(v) = A \cdot v$ earlier, we can now see that what we have above is equivalent to

$$A \cdot v = 0$$

But this is exactly the null space of the matrix A (see Section 6.6).

Note that $A \cdot v = 0$ always had a solution so that the null space was not the empty set. Likewise, the kernel is non-empty as we have shown above that the zero vector of the domain belongs to it.

As the null space was a vector subspace of $\Re^n$ for some n, so the kernel of T is a subspace of the domain V.

Proof

First, the kernel is non-empty since it contains 0_v. Second, we have to show that, given v_1 and v_2 in Ker(T), then $v_1 + v_2 \in$ Ker(T). Now if v_1 and v_2 are in the kernel, then, by definition,

$$T(v_1) = T(v_2) = 0_w$$

Now,

$$T(\mathbf{v}_1 + \mathbf{v}_2) = T(\mathbf{v}_1) + T(\mathbf{v}_2) = \mathbf{0}_w + \mathbf{0}_w = \mathbf{0}_w$$

by the first axiom of a linear transformation and hence

$$\mathbf{v}_1 + \mathbf{v}_2 \in \mathrm{Ker}(T)$$

Finally

$$T(\alpha\mathbf{v}) = \alpha T(\mathbf{v}) = \alpha\mathbf{0}_w = \mathbf{0}_w$$

by the second axiom of a linear transformation and hence $\alpha\mathbf{v} \in \mathrm{Ker}(T)$ whenever $\mathbf{v} \in \mathrm{Ker}(T)$.

Thus we have shown that $\mathrm{Ker}(T)$ is a subspace of V, the domain of the linear transformation.

Example 6H

Let $T: \Re^2 \to \Re^2$ be a linear transformation given by the formula

$$T([x, y]) = [x, 0]$$

Find the kernel of T and give a basis for it.

Solution

We begin by writing the formula in terms of a 2×2 matrix A such that

$$A \cdot \begin{bmatrix} x \\ y \end{bmatrix} = \begin{bmatrix} x \\ 0 \end{bmatrix}$$

It should not be too difficult to fill in the entries of A. We have

$$\begin{bmatrix} 1 & 0 \\ 0 & 0 \end{bmatrix} \begin{bmatrix} x \\ y \end{bmatrix} = \begin{bmatrix} x \\ 0 \end{bmatrix}$$

(With practice you should be able to re-write a transformation T in terms of an appropriate matrix and, vice versa, given a matrix which represents a linear transformation, T, you should be able to display a formula for T.)

What we require is $\text{Ker}(T) = \{\mathbf{v} \in \mathfrak{R}^2 : T(\mathbf{v}) = \mathbf{0}\}$. This is equivalent to the set

$\{[x, y] : [x, 0] = [0, 0]\}$

This means that $x = 0$ and y can be any real number. Thus we have

$\text{Ker}(T) = \{[0, y] \text{ for any } y \in \mathfrak{R}\}$

We can test such elements by mapping them by T and seeing whether or not they have the image, the zero vector of the codomain.

Alternatively, we could have used the matrix equation

$$\begin{bmatrix} 1 & 0 \\ 0 & 0 \end{bmatrix} \begin{bmatrix} x \\ y \end{bmatrix} = \begin{bmatrix} x \\ 0 \end{bmatrix}$$

and ask, what is the null space of the coefficient matrix? This means that we would have to solve the homogeneous equations

$$\begin{bmatrix} 1 & 0 \\ 0 & 0 \end{bmatrix} \begin{bmatrix} x \\ y \end{bmatrix} = \begin{bmatrix} 0 \\ 0 \end{bmatrix}$$

These equations, simply written, would be

$x = 0 \quad \text{and} \quad 0 = 0$

Hence the result would be the same, the null space consisting of vectors of the form $[0 \ \ y]$.

In either case the kernel is one-dimensional since it is spanned by vectors of the form $[0 \ \ y]$. A basis for $\text{Ker}(T)$ could be $\{[0, 1]\}$.

This particular example is a little too simple to need DERIVE. The following example shows where DERIVE might save us some work.

Example 6I

Let T: $\mathfrak{R}^4 \to \mathfrak{R}^3$ be a linear transformation given by the formula

$T([x, y, z, t]) = [x - y + z + t, \ x + 2z - t, \ x + y + 3z - 3t]$

Find the kernel of T and a basis for it.

Solution

First, we observe that the equivalent matrix will be of order 3×4. Second, by definition, the kernel of T is the set of vectors

$$\{[x,\, y,\, z,\, t] = \mathbf{v} \in \mathfrak{R}^4 : T(\mathbf{v}) = \mathbf{0} \in \mathfrak{R}^3\}$$

The equivalent system of homogeneous equations is

$$x - y + z + t = 0$$

$$x + 2z - t = 0$$

$$x + y + 3z - 3t = 0$$

To solve these, we use row-reduction methods. The augmented matrix from this system is

$$\begin{bmatrix} 1 & -1 & 1 & 1 & 0 \\ 1 & 0 & 2 & -1 & 0 \\ 1 & 1 & 3 & -3 & 0 \end{bmatrix}$$

Using DERIVE or by hand, the row-reduced matrix is

$$\begin{bmatrix} 1 & 0 & 2 & -1 & 0 \\ 0 & 1 & 1 & -2 & 0 \\ 0 & 0 & 0 & 0 & 0 \end{bmatrix}$$

We now interpret this result by using the ideas given in Section 6.6.
     First we identify the leading variables as x and y and the free variables as z and t. Thus there are two degrees of freedom, and thus the null space of the matrix has dimension two. This in turn means that the kernel of T has dimension two and thus we are looking for two independent vectors which belong to the kernel, to form a basis for $\text{Ker}(T)$.
     The choices are infinite so we pick two using the method given in Section 6.6. Choose $z = 0$ and $t = 1$ to obtain $y = 2$ and $x = 1$. Thus,

$$[1,\, 2,\, 0,\, 1] \in \text{Ker}(T)$$

Now choose $z = 1$ and $t = 0$ to obtain $y = -1$ and $x = -2$. Thus,

$[-2, -1, 1, 0] \in \text{Ker}(T)$

These two vectors are independent since one is not a multiple of the other so they will do as a basis for $\text{Ker}(T)$. Thus a basis for $\text{Ker}(T)$ could be

$\{[1, 2, 0, 1], [-2, -1, 1, 0]\}$

We now examine the concept of the image of a linear transformation.

Definition

The *image of a linear transformation*, $T: V \rightarrow W$, written $\text{Im}(T)$, is defined by

$\text{Im}(T) = T(V) = \{\mathbf{w} \in W : T(\mathbf{v}) = \mathbf{w} \text{ for some } \mathbf{v} \in V\}$

Alternatively, we could use the definition in Section 6.7 and write

$\text{Im}(T) = T(V) = \{T(\mathbf{v}) : \mathbf{v} \in V\}$

The two definitions are equivalent; it is just a matter of terminology and notation. The first definition stresses the fact that the elements $T(\mathbf{v})$ are vectors in W the codomain. This leads to the fact that, $T(V) \subseteq W$. This means that the direct image of V is a subset of W as shown in Section 6.7. However we now ask the question,

Is $T(V)$ a subspace of W?

The answer is yes, and we will now prove it.

Proof

Recall that we only need to prove three axioms as given in Section 6.3. First, the set $T(V)$ or $\text{Im}(T)$ must contain the zero vector of W. This is a fact since $T(\mathbf{0}_v) = \mathbf{0}_w$.
　　　　Second, we have to show that $\text{Im}(T)$ is closed under vector addition and under scalar multiplication. For vector addition consider two images

$\mathbf{w}_1 = T(\mathbf{v}_1) \in \text{Im}(T)$　　and　　$\mathbf{w}_2 = T(\mathbf{v}_2) \in \text{Im}(T)$

We need to show that

$\mathbf{w}_1 + \mathbf{w}_2 = T(\mathbf{v}_1) + T(\mathbf{v}_2) \in \text{Im}(T)$

Since T is a linear transformation we have

$$\mathbf{w}_1 + \mathbf{w}_2 = T(\mathbf{v}_1) + T(\mathbf{v}_2) = T(\mathbf{v}_1 + \mathbf{v}_2)$$

But since $\mathbf{v}_1 + \mathbf{v}_2 \in V$, because V is a vector space and is closed under addition, it follows that $T(\mathbf{v}_1 + \mathbf{v}_2)$ is an image in W and thus in $\mathrm{Im}(T)$. Hence the second axiom is satisfied.

Finally, we need to show that $\mathrm{Im}(T)$ is closed under scalar multiplication, that is, we need to show that, given any $T(\mathbf{v}) \in \mathrm{Im}(T)$, then $\alpha T(\mathbf{v}) \in \mathrm{Im}(T)$. This again follows quickly from the fact that T is a linear transformation from which we have that

$$\alpha T(\mathbf{v}) = T(\alpha \mathbf{v}) \text{ for all } \mathbf{v} \in V \text{ and scalars } \alpha$$

Hence we have to show that $T(\alpha \mathbf{v}) \in \mathrm{Im}(T)$.

This is true from the fact that since $\mathbf{v} \in V$, $\alpha \mathbf{v} \in V$, because V is a vector space. Thus we have verified all the three axioms for a vector subspace and thus the image of a linear transformation is a subspace of the codomain. We now have in summary that

the kernel of a linear transformation is a subspace of the domain

and

the image of a linear transformation is a subspace of the codomain

We now consider the matrix equivalence of image. To motivate this idea we use an example. Consider the matrix

$$A = \begin{bmatrix} 1 & 2 & 1 \\ 0 & 1 & -1 \\ 2 & 3 & 5 \end{bmatrix}$$

As seen earlier, this can represent a linear transformation T: $\mathfrak{R}^3 \rightarrow \mathfrak{R}^3$, given by $T(\mathbf{v}) = A \cdot \mathbf{v}$ written in appropriate notation.

If we take as our basis for the domain the standard basis for $\mathfrak{R}^3$, we have

$$\{(1, 0, 0), (0, 1, 0), (0, 0, 1)\}$$

If we now map these vectors into the codomain we have the three images

$$T((1, 0, 0)) = A \cdot [1 \ 0 \ 0]^T = [1 \ 0 \ 2]^T$$

$$T((0, 1, 0)) = A \cdot [0 \ 1 \ 0]^T = [2 \ 1 \ 3]^T$$

$$T((0, 0, 1)) = A \cdot [0 \ 0 \ 1]^T = [1 \ -1 \ 5]^T$$

We observe that the images are simply the columns of the matrix A. This will always be the case if we use the standard basis for the domain. We saw this idea in Exercise 6D earlier.

We next ask whether these three images will span $\text{Im}(T)$ and whether they will be independent. In other words, how close are we to getting a basis for $\text{Im}(T)$? We can answer these questions from our matrix work and from our linear transformation work.

From our observation above, the column rank of A will tell us whether or not the three image vectors are independent. If the rank is 3 then the three image vectors will be independent. If the column rank is less than 3, then the vectors will be dependent. Remember that the row rank and column rank are equal.

(Check that for matrix A the row rank is 3.)

Hence the set

$$\{(1, 0, 2), (2, 1, 3), (1, -1, 5)\}$$

is a basis for $\text{Im}(T)$. This also shows that $\text{Im}(T) = \Re^3 = $ codomain of T. Will this always be the case? The answer is, no, as we will see later in further examples.

However, the three image vectors will span $\text{Im}(T)$, as we show in the proposition below for a linear transformation from $\Re^3$ to $\Re^3$. This idea generalizes to a transformation from $\Re^n$ to $\Re^m$.

Proposition

Let $\{e_1, e_2, e_3\}$ be a basis for $V = \Re^3$; then $\{T(e_1), T(e_2), T(e_3)\}$ will span $W = \Re^3$.

Proof

Consider any vector $\mathbf{v}$, in V. It must be expressible in terms of a basis for V and thus we can write

$$\mathbf{v} = \alpha_1 e_1 + \alpha_2 e_2 + \alpha_3 e_3$$

Now,

$$T(\mathbf{v}) = T(\alpha_1 e_1 + \alpha_2 e_2 + \alpha_3 e_3) = \alpha_1 T(e_1) + \alpha_2 T(e_2) + \alpha_3 T(e_3)$$

since T is a linear transformation. This can be interpreted as an element $T(\mathbf{v})$ in Im(T) is a linear combination of the images of our standard basis. Thus any image is simply a linear combination of the columns of A.

As the column rank gives us the dimension of the column space and this is equal to the row rank we have the following fact without further proof:

The dimension of Im(T) is the rank of the matrix A

Thus we have shown that the images obtained above will always span Im(T) but they will not always be independent and therefore they will not always be a basis for Im(T). Consider the matrix

$$A = \begin{bmatrix} 1 & 4 \\ 2 & 8 \end{bmatrix}$$

Let us follow similar steps to those in the example above.

The matrix A will represent a linear transformation $T : \mathfrak{R}^2 \to \mathfrak{R}^2$.

The images of the standard basis

$$\{(1, 0), (0, 1)\}$$

will be

$$\{(1, 2), (4, 8)\}$$

In this case the row rank of A is 1 and therefore the dimension of the image of T is also 1. Observe that one vector is a multiple of the other and thus they form a dependent set. They span the image of T. We can take from our spanning set the vector, $(1, 2)$ as a basis.

As a third example consider the matrix

$$A = \begin{bmatrix} 1 & 2 \\ 3 & 4 \end{bmatrix}$$

We can again define a linear transformation $T: \mathfrak{R}^2 \to \mathfrak{R}^2$ by $T(\mathbf{v}) = A \cdot \mathbf{v}$ where $\mathbf{v}$ is written as a column vector.

(i)   Find the matrix T relative to the basis $\{(1,0), (0,1)\}$. We proceed as we have in the preceding two examples since the basis given is the standard one. The image set of these standard basis vectors is

$$\{(1, 3), (2, 4)\}$$

that is, the columns of A, and therefore the matrix of T, is in fact A.

(ii)   Now find the matrix of T relative to $\{(1,3), (2,5)\}$. We see what happens to the basis vectors $(1, 3)$ and $(2, 5)$ of the domain by working out

$$\begin{bmatrix} 1 & 2 \\ 3 & 4 \end{bmatrix}\begin{bmatrix} 1 \\ 3 \end{bmatrix} \quad \text{and} \quad \begin{bmatrix} 1 & 2 \\ 3 & 4 \end{bmatrix}\begin{bmatrix} 2 \\ 5 \end{bmatrix}$$

In this case we have the image set

$$\{(7, 15), (12, 26)\}$$

Now we express these image vectors in terms of the given basis. This means that we need to solve the equations

$$(7, 15) = \alpha(1, 3) + \beta(2, 5)$$

and

$$(12, 26) = \gamma(1, 3) + \delta(2, 5)$$

Solving these gives $\alpha = -5$, $\beta = 6$, $\gamma = -8$ and $\delta = 10$.
   Thus we have the two equations

$$(7, 15) = -5(1, 3) + 6(2, 5)$$

and

$$(12, 26) = -8(1, 3) + 10(2, 5)$$

We say that the *matrix of T relative to the given basis* is

$$\begin{bmatrix} -5 & -8 \\ 6 & 10 \end{bmatrix}$$

We will expand on this idea in the next section.

6.9   CHANGE OF BASIS

We now use the examples given above to develop some new ideas and concepts.
We have seen that a matrix can represent a linear transformation and vice versa.
          A question we can pose following the last example is, what happens to the
matrix representation when we change from one basis to another?

Let V be a vector space of dimension n.
Let $E = \{\mathbf{e}_1, \mathbf{e}_2, \mathbf{e}_3, ..., \mathbf{e}_n\}$ be a basis for V.
Let $F = \{\mathbf{f}_1, \mathbf{f}_2, \mathbf{f}_3, ..., \mathbf{f}_n\}$ be another basis for V.

By definition we can express each of the $\mathbf{f}_i$s in terms of the $\mathbf{e}_i$s and vice versa. This
means that we can find scalars, a_{ij} for which the following equations hold:

$$\mathbf{f}_1 = a_{11}\mathbf{e}_1 + a_{12}\mathbf{e}_2 + ... + a_{1n}\mathbf{e}_n$$
$$\mathbf{f}_2 = a_{21}\mathbf{e}_1 + a_{22}\mathbf{e}_2 + ... + a_{2n}\mathbf{e}_n$$
$$\mathbf{f}_3 = a_{31}\mathbf{e}_1 + a_{32}\mathbf{e}_2 + ... + a_{3n}\mathbf{e}_n$$

$$\mathbf{f}_n = a_{n1}\mathbf{e}_1 + a_{n2}\mathbf{e}_2 + ... + a_{nn}\mathbf{e}_n$$

We observe that the matrix of coefficients is $A = (a_{ij})$.

Definition

The transpose of A is called *the transition matrix* from the basis $E = \{\mathbf{e}_i\}$ to the basis
$F = \{\mathbf{f}_i\}$.
          Note that in what follows we will let P be the symbol for a transition matrix.
          Notice that the rank of P must be n as the basis vectors are independent. It
follows, therefore, that P has an inverse. In fact, P is the transition matrix from the
basis $F = \{\mathbf{f}_i\}$ to the basis $E = \{\mathbf{e}_i\}$. Referring to Problem (ii) in the last section we
had the standard basis as well as the basis $\{(1, 3), (2, 5)\}$ and the basis $\{(7, 15),
(12, 26)\}$.
          The transition matrix from $E = \{(1, 0), (0, 1)\}$ to $F = \{(7, 15), (12, 26)\}$ would
be

$$\begin{bmatrix} 7 & 12 \\ 15 & 26 \end{bmatrix}$$

since

$$(7, 15) = 7(1, 0) + 15(0, 1)$$

and

$$(12, 26) = 12(1, 0) + 26(0, 1)$$

If we started again with the basis E now equal to $\{(1, 3), (2, 5)\}$ and F the same as above, the transition matrix from E to F would be the one given immediately before this section, that is,

$$\begin{bmatrix} -5 & -8 \\ 6 & 10 \end{bmatrix}$$

since it comes directly from the equations we solved for α, β, γ and δ.

From what we have said above you should be able to write down the transition matrix from $F = \{(7, 15), (12, 26)\}$ back to $E = \{(1, 3), (2, 5)\}$.

The required matrix is

$$\begin{bmatrix} -5 & -8 \\ 6 & 10 \end{bmatrix}^{-1} = -\frac{1}{2}\begin{bmatrix} 10 & 8 \\ -6 & -5 \end{bmatrix} = \begin{bmatrix} -5 & -4 \\ 3 & 5/2 \end{bmatrix}$$

This in turn means that the following relationships are true, as can be checked by direct computation:

$$(1, 3) = -5(7, 15) + 3(12, 26)$$

and

$$(2, 5) = -4(7, 15) + \tfrac{5}{2}(12, 26)$$

Before we do an exercise on these ideas we introduce a further piece of notation.

Notation

If $T: V \rightarrow V$ is a linear transformation and V has a basis $E = \{e_i\}$ then $\{T(\mathbf{e}_1) = \mathbf{f}_1,..., T(\mathbf{e}_n) = \mathbf{f}_n\}$ can be expressed in terms of the $\mathbf{e}_i$s as above. In this notation we write $[T]_E$ and talk of the matrix of T relative to E, or simply the matrix of T in the basis E.

Example 6J

Consider the two bases $E = \{(1, 0), (0, 1)\}$ and $F = \{(1, 3), (2, 5)\}$ of $\mathfrak{R}^2$; then the transition matrix, P, from E to F would be given directly by

$$\begin{bmatrix} 1 & 2 \\ 3 & 5 \end{bmatrix}$$

Write down the transition matrix from F to E.

Solution

Choose any vector $\mathbf{v}$ in $\mathfrak{R}^2$, and show that this vector expressed in the basis F is equal to the inverse of the transition matrix above, multiplied by the same vector expressed in the basis E. In symbols we would write

$$[\mathbf{v}]_F = P^{-1} \cdot [\mathbf{v}]_E$$

As an example we will choose $\mathbf{v} = (-3, 2)$.

  We now need to express this vector in basis E and F to obtain $[\mathbf{v}]_E$, $[\mathbf{v}]_F$ respectively. This means that we need to find α and β such that

$$(-3, 2) = \alpha(1, 0) + \beta(0, 1)$$

and γ and δ such that

$$(-3, 2) = \gamma(1, 3) + \delta(2, 5)$$

From the first equation we can write down $\alpha = -3$ and $\beta = 2$. From the second equation we have to work a little harder to obtain $\gamma = 19$ and $\delta = -11$. Hence we write

$$(-3, 2)_E = (-3, 2)$$

and

$$(-3, 2)_F = (19, -11)$$

This means that the vector $\mathbf{v} = (-3, 2)$ when expressed in the standard basis is exactly the same, but when $\mathbf{v}$ is expressed in terms of the basis F it needs 19 lots of $\mathbf{f}_1$ and -11 lots of $\mathbf{f}_2$ to generate it.

  We will now verify the equation

$$[\mathbf{v}]_F = P^{-1} \cdot [\mathbf{v}]_E$$

Now the right-hand side of this equation is

$$\begin{bmatrix} -5 & 2 \\ 3 & -1 \end{bmatrix} \cdot \begin{bmatrix} -3 \\ 2 \end{bmatrix} = \begin{bmatrix} 19 \\ -11 \end{bmatrix}$$

We observe that the product is indeed $[\mathbf{v}]_F$.

You are urged to try your own choice of vector $\mathbf{v}$ and show that the result

$$[\mathbf{v}]_F = P^{-1} \cdot [\mathbf{v}]_E$$

still holds.   You could attempt to work in general terms to prove the general statement.   In this case start with the vector $\mathbf{v} = (x, y)$.

Example 6K

Let P be the transition matrix from a basis E to a basis F in a vector space V, and let $T: V \rightarrow V$ be any linear transformation; then

$$[T]_F = P^{-1}[T]_E P$$

Proof

The Trivial Case

We will choose the two bases E and F given at the start of Example 6J.   Hence the transition matrix P is as given above.

If we now choose the linear transformation $T: \mathfrak{R}^2 \rightarrow \mathfrak{R}^2$ defined by

$$T((x, y)) = (x + 2y, 3x + 5y)$$

we observe that the matrix T relative to the basis E, $[T]_E$, is again P.   Further, if we now work out the matrix of T relative to the basis F, $[T]_F$ it will also turn out to be P. Verify each of these statements for yourself.

Hence the relation

$$[T]_F = P^{-1}[T]_E P$$

is trivially satisfied.

A Non-trivial Case

Again let E and F be as before so that P is as before. This time let us choose the linear transformation $T: \Re^2 \to \Re^2$ given by

$$T((x, y)) = (2y, 3x - y)$$

In this case,

$$[T]_E = \begin{bmatrix} 0 & 2 \\ 3 & -1 \end{bmatrix}$$

as can be verified directly. Furthermore $T((1, 3)) = (6, 0)$ and $T((2, 5)) = (10, 1)$. Now we can work out $P^{-1}[T]_E P$.

In full it is

$$\begin{bmatrix} -5 & 2 \\ 3 & -1 \end{bmatrix}\begin{bmatrix} 0 & 2 \\ 3 & -1 \end{bmatrix}\begin{bmatrix} 1 & 2 \\ 3 & 5 \end{bmatrix} = \begin{bmatrix} -30 & -48 \\ 18 & 29 \end{bmatrix}$$

Is this the same as $[T]_F$? If it is then it should mean that the following equations are true:

$$(6, 0) = -30(1, 3) + 18(2, 5)$$

and

$$(10, 1) = -48(1, 3) + 29(2, 5)$$

By simple arithmetic we can verify that the equations are correct. This has saved us from computing $[T]_F$ directly.

6.10   SIMILARITY

Definition

Two square matrices A and B of the same size are said to *be similar* if there exists a matrix P which has an inverse P^{-1} such that

$$B = P^{-1}AP$$

This relationship defines an equivalence relation. We have not defined this in this book but the idea can be found in any standard text on relations and functions. It is not important at this stage that you understand this idea.

We say that A and B are equivalent under this equivalence relation. Thus we can say that two matrices A and B represent the same linear transformation if and only if A and B are similar.

We had an example of this at the end of the last section. There, $[T]_E$ and $[T]_F$ were shown to be similar and a certain transition matrix P was obtained which always had an inverse. We now move towards generalizing what we achieved in the last section by means of a DERIVE activity.

DERIVE ACTIVITY 6D

(A)   Let $A = \begin{bmatrix} 2 & 1 \\ 0 & -1 \end{bmatrix}$ and $B = \begin{bmatrix} 4 & -2 \\ 5 & -3 \end{bmatrix}$.

Are these matrices similar?
This means: is there a matrix

$$P = \begin{bmatrix} a & b \\ c & d \end{bmatrix}$$

which has an inverse and is such that $P^{-1}AP = B$? An alternative way of looking at this expression is to multiply both sides of the equation by P so that we obtain

$$AP = PB$$

Sometimes it is more convenient to work with this alternative equation. It could save us from working out P^{-1} directly.

Investigate, using DERIVE, what P might be by trying to answer the questions

(i)   Does P exist with this property?
(ii)   If so, is P unique?

Do this before reading on.
If we work out AP and PB we obtain the following matrices:

$$AP = \begin{bmatrix} 2a+c & 2b+d \\ -c & -d \end{bmatrix}$$

and

$$PB = \begin{bmatrix} 4a + 5b & -2a - 3b \\ 4c + 5d & -2c - 3d \end{bmatrix}.$$

Putting $AP = PB$ we obtain four equations by equating elements. However, we only obtain two different equations and they are

$$d = -c$$

and

$$c = 2a + 5b$$

Thus P has to have the form

$$\begin{bmatrix} a & b \\ 2a + 5b & -(2a + 5b) \end{bmatrix}$$

Furthermore, for P to have an inverse $\text{DET}(P) \neq 0$:

$$\text{DET}(P) = -(2a + 5b) \begin{vmatrix} a & b \\ -1 & 1 \end{vmatrix} = -(2a + 5b)(a + b)$$

Hence we must choose a and b such that $a + b \neq 0$ and $(2a + 5b) \neq 0$. Such a choice could be $a = 2$ and $b = -1$. Then

$$P = \begin{bmatrix} 2 & -1 \\ -1 & 1 \end{bmatrix}$$

Note that $\text{DET}(P) = 1$ and thus has an inverse.
　　　Check directly that $AP = PB$.
　　　Choose other values for a and b which satisfy the criteria above and check again that your matrix P satisfies the equation $AP = PB$.
　　　We have thus answered the questions posed earlier:

(i)　　P does exist such that $AP = PB$, but we have to choose carefully.
(ii)　　P is not unique.

(B)   Consider the diagonal matrix $D = \begin{bmatrix} 2 & 0 \\ 0 & -1 \end{bmatrix}$ and the matrix B given in (A).

The question is the same as above:

Are the matrices D and B similar?

Using a similar approach to the one given above, investigate this question before reading on.

In this case $DP = PB$ in general terms gives

$$\begin{bmatrix} 2a & 2b \\ -c & -d \end{bmatrix} = \begin{bmatrix} 4a+5b & -2a-3b \\ 4c+5d & -2c-3d \end{bmatrix}$$

Equating elements we obtain two essentially different equations, namely,

$$c + d = 0 \quad \text{and} \quad 2a + 5b = 0$$

Thus P must be of the form

$$\begin{bmatrix} a & -2a/5 \\ c & -c \end{bmatrix}$$

As before, the determinant of P must be non-zero, so that

$$-3ac/5 \neq 0$$

This implies that $a \neq 0$ and $c \neq 0$.

Choose a and c to satisfy these criteria, for example $a = 5$ and $c = 1$ giving

$$P = \begin{bmatrix} 5 & -2 \\ 1 & -1 \end{bmatrix}$$

Hence we can produce any number of matrices for P.

What we have achieved in this activity is that the matrix B can be made similar to a diagonal matrix, D.

Definition

A linear transformation T is said to be *diagonalizable* if for some basis it can be represented by a diagonal matrix.

It follows from this that if A is the matrix representation of a linear transformation T, then T is diagonalizable if and only if there exists a matrix P with an inverse P^{-1} such that $P^{-1}AP$ is a diagonal matrix.

Looking again at DERIVE Activity 6D, we can associate with matrix B given there a linear transformation $T: \mathfrak{R}^2 \to \mathfrak{R}^2$ defined by

$$T((x,\ y)) = (4x - 2y,\ 5x - 3y)$$

If P is given by

$$P = \begin{bmatrix} 5 & -2 \\ 1 & -1 \end{bmatrix}$$

it is clear that P^{-1} exists as DET(P) = -3. We do not need to compute P^{-1} explicitly since we know from the DERIVE Activity 6D, that $DP = PB$ and hence $B = P^{-1}DP$. This means that B is similar to D and hence T is diagonalizable.

The questions we might ask at this stage are:

(i)   How do we find a matrix P in the kind of problems we solved above, without having to start from the beginning with a general matrix?

(ii) Are all matrices/transformations diagonalizable?

We answer the first question in the next chapter where we study eigenvalues and eigenvectors. We answer the second question now:

Not all $n \times n$ matrices are diagonalizable

Consider, for example, the matrix

$$\begin{bmatrix} 2 & 1 \\ 0 & 2 \end{bmatrix}$$

Is this matrix similar to a diagonal matrix of the form

$$\begin{bmatrix} x & 0 \\ 0 & y \end{bmatrix}$$

or alternatively does there exist a matrix P which has an inverse with the property that

$$\begin{bmatrix} 2 & 1 \\ 0 & 2 \end{bmatrix} \cdot P = P \cdot \begin{bmatrix} x & 0 \\ 0 & y \end{bmatrix}?$$

If we take P as the general matrix:

$$\begin{bmatrix} a & b \\ c & d \end{bmatrix}$$

we have

$$\begin{bmatrix} 2 & 1 \\ 0 & 2 \end{bmatrix}\begin{bmatrix} a & b \\ c & d \end{bmatrix} = \begin{bmatrix} a & b \\ c & d \end{bmatrix}\begin{bmatrix} x & 0 \\ 0 & y \end{bmatrix}$$

from which we arrive at the four equations given below:

$$2a + c = ax \tag{6.3}$$

$$2b + d = by \tag{6.4}$$

$$2c = cx \tag{6.5}$$

$$dy = 2d \tag{6.6}$$

First we observe that $a = b = c = d = 0$ is a solution but then P would be the zero matrix and it doesn't have an inverse. Hence we are looking for a non-trivial solution to the four equations. We show below that there are no such solutions.

From (6.5) either $c = 0$ or $x = 2$.
From (6.6) either $d = 0$ or $y = 2$.

Consider the four cases given below.

Case 1

$c = 0$ and $d = 0$ ($x \neq 2$, $y \neq 2$).
    From (6.3) $2a = ax$ and thus as $x \neq 2$ we must conclude that $a = 0$. From (6.4) $2b = by$ and thus as $y \neq 2$ we must conclude that $b = 0$. Hence we have for P the zero matrix, which is not acceptable.

Case 2

$c = 0$ and $d \neq 0$ ($x \neq 2$, $y = 2$). From (6.4) $2b + d = 2b$ and thus $d = 0$, which contradicts our assumption. Hence we have no matrix for P in this case.

Case 3

$c \neq 0$ and $d = 0$ ($x = 2$, $y \neq 2$). From (6.3) $2a + c = 2a$ and thus $c = 0$, which contradicts our assumption. Hence we have no matrix for P in this case.

Case 4

$c \neq 0$ and $d \neq 0$ ($x = 2$, $y = 2$). From either (6.3) or (6.4) we have the same contradiction as in cases 2 and 3. Hence we conclude that we cannot find a non-trivial matrix for P, as claimed. Therefore, we have found a matrix, namely,

$$\begin{bmatrix} 2 & 1 \\ 0 & 2 \end{bmatrix}$$

which is not diagonalizable.

We finish this chapter by considering the trace of a matrix and some of its properties.

Definition

The *trace* of a square matrix, denoted by $\text{Tr}(A)$, is the sum of its diagonal elements. This means that if $A = (a_{ij})$ is an $n \times n$ matrix then $\text{Tr}(A) = \sum_{i=1}^{n} a_{ii}$.

DERIVE ACTIVITY 6E

Input any 3×3 square matrix in line 1. Then **Author**

TRACE(#1)

Now **Simplify** to show that your answer is the sum of the diagonal elements. Try square matrices of other sizes to familiarize yourself with this new function. By choosing suitable matrices show that

(i) $\text{Tr}(AB) = \text{Tr}(BA)$
        and
(ii) $\text{Tr}(A) = \text{Tr}(P^{-1}AP)$

This last result means that similar matrices have the same trace. Look back at DERIVE Activity 6D and check this result for the matrices A and B given there.
   We will now prove these two results, below, in general:

(i) $\text{Tr}(AB) = \text{Tr}(BA)$.
         Let $A = (a_{ij})$ and $B = (b_{ij})$ be two $n \times n$ matrices. Then $AB = C = (c_{ij})$, where $c_{ij} = a_{i1}b_{1j} + a_{i2}b_{2j} + \ldots + a_{in}b_{nj}$ where $i = 1, \ldots, n$ and $j = 1, \ldots, n$. This is the formal definition of matrix multiplication.
         The trace of AB is simply the sum of the elements c_{ii} for $i = 1, \ldots, n$. Thus the trace of AB is

$$\sum_{i=1}^{n} c_{ii} = \sum_{i=1}^{n} a_{i1}b_{1i} + a_{i2}b_{2i} + \ldots + a_{in}b_{ni}$$

Similarly, the trace of BA is

$$\sum_{i=1}^{n} b_{1i}a_{i1} + b_{2i}a_{i2} + \ldots + b_{ni}a_{in}$$

Hence the result follows.

(ii) $\text{Tr}(A) = \text{Tr}(P^{-1}AP)$.
         To prove this statement we use (i) above.
         Consider $\text{Tr}(P^{-1}AP) = \text{Tr}((P^{-1})(AP))$ using associativity of matrices. Now using (i) on this product of P^{-1} and AP we can write

$$\text{Tr}(P^{-1}AP) = \text{Tr}((P^{-1})(AP)) = \text{Tr}((AP)(P^{-1}))$$

Using associativity again we can write

$$\text{Tr}((AP)(P^{-1})) = \text{Tr}(APP^{-1})$$

And, finally, using the inverse axiom that $PP^{-1} = I$, the identity matrix, and the identity axiom that $AI = A$, we have

$$\text{Tr}(APP^{-1}) = \text{Tr}(AI) = \text{Tr}(A)$$

This proves (ii).
         This work on traces will prove particularly useful in the next chapter.

Consolidation Exercise

1.   Consider each of the mappings given below and decide whether or not they are linear transformations:

     (i) $F: \Re^2 \to \Re^2$ defined by $F(x, y) = (x + y, x)$
     (ii) $F: \Re^2 \to \Re$ defined by $F(x, y) = xy$
     (iii) $F: \Re^3 \to \Re$ defined by $F(x, y, z) = 2x - 3y + 4z$
     (iv) $F: \Re^2 \to \Re^3$ defined by $F(x, y) = (x + 1, 2y, x + y)$
     (v) $F: \Re^3 \to \Re^2$ defined by $F(x, y, z) = (|x|, 0)$

     Try to find elegant solutions to these five questions by recalling the properties of linear transformations and by considering such things as counterexamples.

2.   Prove that the mappings f and g given below are linear transformations:

$$f: \Re^2 \to \Re^3 \text{ defined by } f(x, y) = (x + y, x + 2y, 3x)$$

and

$$g: \Re^3 \to \Re^2 \text{ defined by } g(x, y, z) = (x + y + z, 2x + 3z)$$

     Give a formula for, $(g \circ f)$, the composition of the two functions. Write down the matrices which represent f, g and $(g \circ f)$ and explain how they are related.
           Work out the kernel and image for each of the transformations

$$f, g \text{ and } (g \circ f)$$

     Verify in each case, the dimension theorem for linear transformations which states

     The dimension of the domain   = the dimension of the kernel
                                            + the dimension of the image.

3.   Compute the dimension of the vector spaces spanned by the following sets:
     (i) $\{(1, -2, 3, -1), (1, 1, -2, 3)\}$
     (ii) $\{t^3 + 2t^2 + 3t + 1, 2t^3 + 4t^2 + 6t + 2\}$
     (iii) $\{t^3 - 2t^2 + 5, t^2 + 3t - 4\}$
     (iv) $\left\{ \begin{bmatrix} 1 & 2 \\ 1 & 2 \end{bmatrix} \begin{bmatrix} 1 & 1 \\ 2 & 2 \end{bmatrix} \right\}$

4.   Find a basis and the dimension of both the image and kernel of the following
     linear transformations:

  (i) $F: \Re^3 \to \Re$ defined by $F(x, y, z) = 2x - 3y + 5z$
  (ii) $F: \Re^3 \to \Re^2$ defined by $F(x, y, z) = (x + y, y + z)$

5.   Let V be the Euclidean vector space $\Re^3$.
       Show that the set U is a vector subspace of V where

     $U = \{(x, y, z) : x + y + z = 0\}$

     For example, the vectors $(1, 1, -2)$ and $(0, 1, -1)$ belong to U.

6.   Let V be the vector space of all 2×2 matrices with real entries and let the
     scalars be the real numbers, $\Re$.

  (i)   Show that the subset U of V which consists of all matrices whose
         determinant is zero, is not a subspace of V.
             Hint.   Two elements of U are the matrices

$$\begin{bmatrix} 1 & 0 \\ 0 & 0 \end{bmatrix} \text{ and } \begin{bmatrix} 0 & 0 \\ 0 & 1 \end{bmatrix}$$

  (ii)   Is the subset U of V which consists of matrices which are idempotent a
         subspace of V?
             Reminder and hint.   A matrix A which is idempotent is one
         which satisfies $A^2 = A$, and the identity matrix is an example.

7.   Consider the two matrices given below:

$$\begin{bmatrix} 1 & 2 & -1 & 3 \\ 2 & 4 & 1 & -2 \\ 3 & 6 & 3 & -7 \end{bmatrix} \text{ and } \begin{bmatrix} 1 & 2 & -4 & 11 \\ 2 & 4 & -5 & 14 \end{bmatrix}$$

     Decide whether or not these two matrices have the same row space.

8.   Consider the standard basis for the Euclidean vector space $\Re^3$, call it E, and
     the basis $F = \{(1, 1, 1), (1, 1, 0), (1, 0, 0)\}$:

  (i)   Write down the transition matrix, P, from E to F.
  (ii)   Work out the transition matrix, Q, from F to E.
  (iii) Verify that $PQ = I$.

9.   Show that similarity of matrices is an equivalence relation on the set of $n \times n$ matrices. Let A, B and C be three $n \times n$ matrices.

          The above problem means that you have to show that the following three statements are true:

    (i) A is similar to itself.
    (ii)   If A is similar to B then B is similar to A.
    (iii)   If A is similar to B and B is similar to C, then A is similar to C.

7

Eigenvalues and Eigenvectors

7.1   EIGENVALUES AND EIGENVECTORS

We motivate this topic by introducing some simple problems for the reader to attempt before looking at our solutions.

Example 7A

If $B = \begin{bmatrix} 1 & 3 \\ 5 & 3 \end{bmatrix}$ and $f(x) = 2x^2 - 4x + 3$, find $f(B)$.

Comment

Here we have a 2×2 matrix and a polynomial for the function f. They seem to have no relation to each other until we say let x be the matrix B. The counterpart to the polynomial in x will now look as

$$f(B) = 2B^2 - 4B + 3I$$

We have to put $3I$ rather than just 3 for conformability of matrices under addition and subtraction. Remember, I is the appropriate identity matrix. From Chapter 1 we know what B^2 means and we know what a scalar times a matrix is, so we can now give the solution.

Solution

Using DERIVE or by hand we work out B^2 to be

$$\begin{bmatrix} 16 & 12 \\ 20 & 24 \end{bmatrix}$$

Thus,

$$f(B) = 2B^2 - 4B + 3I$$

$$= \begin{bmatrix} 32 & 24 \\ 40 & 48 \end{bmatrix} - \begin{bmatrix} 4 & 12 \\ 20 & 12 \end{bmatrix} + \begin{bmatrix} 3 & 0 \\ 0 & 3 \end{bmatrix}$$

$$= \begin{bmatrix} 31 & 12 \\ 20 & 39 \end{bmatrix}$$

Example 7B

In this problem use the same matrix B as in Example 7A, above. The function g, is given by the formula $g(x) = x^2 - 4x - 12$. Find $g(B)$.

Solution

Here,

$$g(B) = \begin{bmatrix} 16 & 12 \\ 20 & 24 \end{bmatrix} - \begin{bmatrix} 4 & 12 \\ 20 & 12 \end{bmatrix} + \begin{bmatrix} 12 & 0 \\ 0 & 12 \end{bmatrix}$$

$$= \begin{bmatrix} 0 & 0 \\ 0 & 0 \end{bmatrix}$$

Comment

In this case we have the zero matrix for $g(B)$. Using the same terminology as in equations, we say that B satisfies the polynomial equation $g(x) = 0$. Alternatively, we say that B is a solution of the equation $g(x) = 0$.

Example 7C

Again use the same matrix B. Find a non-zero column vector $\mathbf{u} = \begin{bmatrix} x \\ y \end{bmatrix}$ such that

$B\mathbf{u} = 6\mathbf{u}$.

Solution

We have $B\mathbf{u} = 6\mathbf{u}$ equivalent to

$$\begin{bmatrix} 1 & 3 \\ 5 & 3 \end{bmatrix}\begin{bmatrix} x \\ y \end{bmatrix} = \begin{bmatrix} 6x \\ 6y \end{bmatrix}$$

Multiplying out the left-hand side and equating the result with the right-hand side we have the two equations

$$x + 3y = 6x$$

and

$$5x + 3y = 6y$$

However, these two reduce to the same equation, namely,

$$5x - 3y = 0$$

From preceding work we can say there is no unique solution.

Furthermore, we have an infinity of solutions or a line of solutions. We have one degree of freedom so we can choose either x or y and the other variable will be defined.

Choose $x = 3a$, then $y = 5a$. Thus we have an infinity of solutions of the form

$$\begin{bmatrix} 3a \\ 5a \end{bmatrix}$$

where $a \neq 0$, since a non-zero solution is required.

We will now use these three problems to introduce the idea of eigenvalues and eigenvectors. For example, in Example 7C we found a family of vectors which satisfied a relation of the form

$$B\mathbf{x} = 6\mathbf{x}$$

In general we ask, can we find a column $n \times 1$ vector $\mathbf{x}$ which satisfies the matrix equation

$$A\mathbf{x} = \lambda\mathbf{x}$$

where λ is a scalar and A is an $n \times n$ matrix? If the answer is yes, then we say that λ is the *eigenvalue of A corresponding to the eigenvector* **x** *of A*. In Example 7C, we found that B had an eigenvalue 6 and the corresponding eigenvector was $u = [3a \;\; 5a]^{T}$.

How do we find the eigenvalues without being told them as in Example 7C?

Once we have an eigenvalue we can work out the eigenvector in a similar way to Example 7C.

We answer the question just posed by working in general.

Let A be an $n \times n$ matrix, **x** be an $n \times 1$ column vector and λ a scalar. We now consider the matrix equation

$$A\mathbf{x} = \lambda \mathbf{x}$$

This can be rewritten as

$$A\mathbf{x} - \lambda \mathbf{x} = \mathbf{0}$$

where **0** is the $n \times 1$ zero matrix. We can further write this last equation as

$$(A - \lambda I)\mathbf{x} = \mathbf{0}$$

Recall the comment of Example 7A that when a scalar is on its own we need to introduce the identity matrix to make all the operations conform.

The matrix equation

$$(A - \lambda I)\mathbf{x} = 0$$

is a homogeneous system of equations in n unknowns, the entries of **x**. From our preceding work in Chapter 5, Section 5.3, we know that this system has a trivial solution, $\mathbf{x} = \mathbf{0}$. However, we require a non-trivial solution, that is, a non-zero solution. Again from our preceding work this means that we want the determinant of the matrix of coefficients to be zero, that is,

$$\text{DET}(A - \lambda I) = 0$$

This follows from the statement in Section 5.2 which states that for an infinity of solutions $\text{Rank}(A) < n$, and, furthermore, $\text{Rank}(A|0) = \text{Rank}(A)$.

Take as an example at this stage, the matrix B from Example 7A. Here $B - \lambda I$ is given by

$$\begin{bmatrix} 1-\lambda & 3 \\ 5 & 3-\lambda \end{bmatrix}$$

Now consider the determinant of this matrix:

$$\text{DET}(B - \lambda I) = (1 - \lambda)(3 - \lambda) - 15$$

$$= \lambda^2 - 4\lambda - 12$$

Compare this quadratic expression with the function g given in Example 7B. For a non-trivial solution we require $\text{DET}(B - \lambda I) = 0$ and thus we require

$$\lambda^2 - 4\lambda - 12 = 0$$

Solving for λ we have

$$\lambda^2 - 4\lambda - 12 = (\lambda - 6)(\lambda + 2) = 0$$

which implies that $\lambda = 6$ or $\lambda = -2$. Hence we have two eigenvalues for the matrix B, namely, 6 and –2. Recall that 6 was given as an eigenvalue in Example 7C. To find the corresponding eigenvectors we have to consider two cases as follows.

Case 1

$$\lambda = 6$$

We need to solve $B\mathbf{x} = 6\mathbf{x}$. This is exactly Example 7C. The solution was $\mathbf{x} = [3a \ \ 5a]^T$ with $a \neq 0$. Thus the eigenvector corresponding to the eigenvalue 6 is of this form.

Case 2

$$\lambda = -2$$

We need to solve $B\mathbf{x} = -2\mathbf{x}$. This means that we solve

$$\begin{bmatrix} 1 & 3 \\ 5 & 3 \end{bmatrix} \begin{bmatrix} x \\ y \end{bmatrix} = \begin{bmatrix} -2x \\ -2y \end{bmatrix}$$

This in turn reduces to just one equation, namely,

$$x + y = 0$$

Choose $x = a$ and then $y = -a$. Hence for the eigenvalue –2, the corresponding eigenvector is of the form

$$x = [a \quad -a]^T$$

The problem, which in many books is called *the eigenvalue problem*, is no different for a 3 × 3 matrix and higher-order ones. The work, however, increases dramatically and, once again, this is where DERIVE comes in to help us.

Definition

The equation $DET(A - \lambda I) = 0$ is called *the characteristic equation*

In the above we had the characteristic equation for B as

$$\lambda^2 - 4\lambda - 12 = 0$$

Observe that if A is an $n \times n$ matrix, then the characteristic equation will be a polynomial of the nth degree. We would therefore expect n solutions to the characteristic equation provided we allowed complex solutions. Note that eigenvalues can be complex numbers.

The Cayley–Hamilton Theorem

A square matrix always satisfies its characteristic equation.

Refer back to Example 7B and observe that we showed that this theorem was true in the case of the matrix B.

An alternative statement is that a matrix is always a zero of its characteristic equation. We will not prove this theorem as it can be found in most of the classical texts. The usual proof relies on the adjoint matrix.

Exercise 7A

1.   Consider the matrix

$$A = \begin{bmatrix} 2 & 3 \\ -1 & -2 \end{bmatrix}$$

(i)   Show that the characteristic equation for the eigenvalues is

$$\lambda^2 - 1 = 0$$

and hence find the eigenvalues.

(ii)   Show that the matrix A satisfies its characteristic equation.

(iii) Find the eigenvectors of A.

2.   Repeat the activities of Problem 1 for the matrix

$$C = \begin{bmatrix} 1 & 2 \\ 12 & 3 \end{bmatrix}$$

DERIVE has functions which can find the characteristic polynomial, and the eigenvalues of square matrices.   To find the exact eigenvectors or approximate eigenvectors we need to load the utility file VECTOR.MTH, as we did in an earlier chapter.

The activity which follows will introduce these new functions to the reader.

DERIVE ACTIVITY 7A

(A)   On line 1 input the matrix B of Example 7A.

Author

        CHARPOLY(#1)

Then **Simplify**.

        DERIVE defaults to w which replaces our λ. If you want a particular variable you would type,

        CHARPOLY(#1,x)

In this case you would obtain the same polynomial but with variable x instead of the default, w. In either case the characteristic equation is of the form

$$w^2 - 4w - 12$$

as seen previously. This should appear on your third line.
        Now **Author**

        EIGENVALUES(#1)

In this case you obtain the line

        $[w = -2, w = 6]$

Again if you want a particular variable, say x, you **Author**

        EIGENVALUES(#1,x)

Now load the utility file VECTOR.MTH, **Author**

EXACT_EIGENVECTOR(#1,6)

This statement is asking DERIVE to work out the eigenvector corresponding to the eigenvalue 6. Because we know it is exact this function produces the answer that we already obtained in Example 7C, but in a slightly different form.

Probably on your screen will be something like

$$\left[x1 = @1 \quad x2 = \frac{5@1}{3} \right]$$

This means that if we choose x_1, which is our x, as 'a', then x_2, which is our y, will be $5a/3$.

If we scale these values up by 3 we obtain the solution $[3a \quad 5a]^T$, as before. Remember we are obtaining a line of solutions and therefore there is an infinity of ways of presenting the eigenvector!

Now let us find the eigenvector for the eigenvalue -2.

Author

EXACT_EIGENVECTOR(#1,−2)

and then **Simplify**.

Your next line should be something like

$[x1 = @2 \quad x2 = -@2]$

Again this means that if we choose x_1 to be 'a', then x_2 will be '$-a$'. This agrees exactly with what we worked out in Case 2, above.

(B)   Repeat the activities in (A) for the matrices of Exercise 7A:

$$A = \begin{bmatrix} 2 & 3 \\ -1 & -2 \end{bmatrix} \qquad C = \begin{bmatrix} 1 & 2 \\ 12 & 3 \end{bmatrix}$$

(C)   Investigate what happens for the 3×3 matrix

$$\begin{bmatrix} 1 & 0 & 3 \\ 1 & 1 & 0 \\ 2 & 1 & 4 \end{bmatrix}$$

We now investigate 3×3 matrices algebraically.

Example 7D

Let

$$M = \begin{bmatrix} 1 & 0 & 3 \\ 1 & 1 & 0 \\ 2 & 1 & 4 \end{bmatrix}$$

Show that the characteristic equation of the matrix M is

$$\lambda^3 - 6\lambda^2 + 3\lambda - 1 = 0$$

Using the Cayley–Hamilton theorem compute M^{-1}.

Comments

Of course, we could use DERIVE immediately to find the inverse of M but we are asked to find M^{-1} in a particular way. At least we can check our answer using DERIVE directly!

The variable λ which is commonly used for eigenvalues, is not obtainable in DERIVE but twelve Greek letters are. These are the ones in the extended ASCII character set. An example is given below in the solution to this problem. See the DERIVE manual for further information.

Solution

We will assume that you have loaded the matrix M in line 1, but if not use your correct line number in CHARPOLY below.

To verify the characteristic equation is simply to ask DERIVE

CHARPOLY(#1,alpha)

DERIVE should produce

$$-\alpha^3 + 6\alpha^2 - 3\alpha + 1$$

Hence the characteristic equation is

$$-\alpha^3 + 6\alpha^2 - 3\alpha + 1 = 0$$

which is equivalent to the equation in λ given in the problem statement.

Now from the CayleyHamilton theorem, we know that M satisfies the equation given above. Hence,

$$-M^3 + 6M^2 - 3M + I = 0 \quad \text{or} \quad M^3 - 6M^2 + 3M - I = 0$$

Although we are asked to compute M^{-1} it is a good idea to check that it exists by evaluating DET(M) and showing that it is non-zero. Check this now using DERIVE.

Since M^{-1} exists we can multiply the equation

$$M^3 - 6M^2 + 3M - I = 0$$

throughout by M^{-1} to obtain

$$M^2 - 6M + 3I - M^{-1} = 0$$

Hence we have

$$M^{-1} = M^2 - 6M + 3I = \begin{bmatrix} 4 & 3 & -3 \\ -4 & -2 & 3 \\ -1 & -1 & 1 \end{bmatrix}$$

Using DERIVE check that this expression does indeed produce the inverse of the matrix M.

Exercise 7B

(Do this exercise by hand and check your answers using DERIVE.) For each of the following matrices:

(a)   find the characteristic equation;
(b)   use the Cayley–Hamilton theorem to compute the inverse of each matrix:

$$A = \begin{bmatrix} 5 & 2 \\ 3 & 4 \end{bmatrix} \qquad B = \begin{bmatrix} 4 & 2 & -2 \\ 1 & 3 & 1 \\ -1 & -1 & 5 \end{bmatrix} \qquad C = \begin{bmatrix} -1 & -4 & -6 \\ 1 & 4 & 4 \\ 0 & 0 & 1 \end{bmatrix}$$

We leave this activity to look again at the eigenvalue problem and to interpret it in a geometrical context.

Recall that an $n \times m$ matrix A can be interpreted as a linear transformation $T: \mathfrak{R}^m \to \mathfrak{R}^n$ by the relation $T(\mathbf{v}) = A\mathbf{v}$, where $\mathbf{v}$ is an $m \times 1$ column vector.

Hence if $A\mathbf{x} = \lambda\mathbf{x}$ we can interpret this as a very special transformation that takes the (eigen)vector $\mathbf{x}$ and transforms it to a scalar multiple of itself.

In Example 7C we had the eigenvector $[3a \quad 5a]^T$, where 'a' was any non-zero scalar. We can think of this vector as a continuous line passing through the origin and passing through the points (3, 5), (6, 10), (–3, –5), and so on. In fact, the line given by the equation $3y = 5x$.

So if this equation represents the vector $\mathbf{x}$ what does $6\mathbf{x}$ look like, or $-2\mathbf{x}$ or even $\mathbf{x}/10$? The answer has to be, we can't tell the difference; it still looks like the same infinite line. So an eigenvector is a kind of 'invariant'. It doesn't seem to change under a certain transformation. We now explore this idea through examples, which we have already seen earlier in this book (Section 6.7).

Reflections

In the plane we had the linear transformation which reflected vectors in the y axis; it was given by the matrix

$$\begin{bmatrix} -1 & 0 \\ 0 & 1 \end{bmatrix}$$

Before we work out the eigenvalues and eigenvectors of this matrix consider from a geometrical point of view which vectors (infinite lines) do not essentially change their positions under such a reflection.

Note   We have made the statement plural so there is more than one such line! Think about this question before reading on.

Your first thought might have been the 'mirror line', that is, the y axis. Such a line doesn't 'move'. Did you get a second line?

Let $A = \begin{bmatrix} -1 & 0 \\ 0 & 1 \end{bmatrix}$. Then the characteristic equation is DET$(A - \lambda I) = 0$. This is equivalent to

$$\begin{vmatrix} -1-\lambda & 0 \\ 0 & 1-\lambda \end{vmatrix} = 0$$

This gives

$$(-1 - \lambda)(1 - \lambda) = 0$$

from which we obtain the solutions

$$\lambda = 1 \quad \text{or} \quad \lambda = -1$$

You could verify these stages using DERIVE but because we are choosing simple examples we hope you can do them quickly by hand.

Case 1

$$\lambda = 1$$

In this case the problem we must solve is $A\mathbf{x} = 1\mathbf{x} = \mathbf{x}$. This is equivalent to

$$\begin{bmatrix} -1 & 0 \\ 0 & 1 \end{bmatrix}\begin{bmatrix} x \\ y \end{bmatrix} = \begin{bmatrix} x \\ y \end{bmatrix}$$

This matrix equation gives us two equations, namely,

$$-x = x$$

and

$$y = y$$

The second equation gives us no information about y, whereas the first equation implies that $x = 0$. Hence the eigenvector corresponding to the eigenvalue 1 is of the form

$$\begin{bmatrix} 0 \\ a \end{bmatrix}$$

This in its turn is simply the y axis, the 'mirror line' of this transformation. We can check that vectors of this type do satisfy the equation $A\mathbf{x} = \mathbf{x}$ by direct computation.

Case 2

$$\lambda = -1$$

In this case the problem we must solve is $A\mathbf{x} = -\mathbf{x}$. This is equivalent to

$$\begin{bmatrix} -1 & 0 \\ 0 & 1 \end{bmatrix} \begin{bmatrix} x \\ y \end{bmatrix} = \begin{bmatrix} -x \\ -y \end{bmatrix}$$

from which we obtain the two equations

$$-x = -x \quad \text{and} \quad y = -y$$

Hence $y = 0$ and x is any scalar. The eigenvectors are therefore of the form

$$\begin{bmatrix} a \\ 0 \end{bmatrix}$$

Such vectors lie on the x axis and are in fact reflected in the origin, but we really can't tell the difference between the original and the transformed line. Again, we can check directly that vectors of this kind satisfy the equation, $A\mathbf{x} = -\mathbf{x}$.

We also observe that the eigenvectors in this problem are orthogonal to each other since their dot product is zero.

We now ask a question. Will every transformation matrix have eigenvalues and therefore eigenvectors?

What you have to consider is: Can you think of a transformation that moves every vector away from its original position?

Rotations

In Section 6.7 we considered rotation matrices. In particular let us consider the matrix representing the linear transformation 'a rotation in $\Re^2$ about the origin through 90° anti-clockwise'.

1.   The 2 × 2 matrix representing this linear transformation is

$$\begin{bmatrix} 0 & -1 \\ 1 & 0 \end{bmatrix}$$

Again we will proceed by hand, but if you wish to verify any results using DERIVE, do so. The characteristic equation is given by the determinant

$$\begin{vmatrix} -\lambda & -1 \\ 1 & -\lambda \end{vmatrix} = 0$$

This gives the equivalent equation

$$\lambda^2 + 1 = 0$$

The solutions of this equation are complex numbers, so if we are seeking real eigenvalues, we conclude that there aren't any. This is not surprising if we consider the geometrical transformation that this problem represents. No vector in $\Re^2$ is mapped to itself or any multiple of itself as it turns physically anti-clockwise about the origin through 90°.

However, if we had rotated through 180° we would have obtained real eigenvalues. We look at this in detail below.

Returning to the equation $\lambda^2 + 1 = 0$ we have for eigenvalues

$$\lambda = i \quad \text{or} \quad \lambda = -i$$

where $i^2 = -1$.

We can now proceed to find the corresponding eigenvectors, which we expect to have complex entries.

Case 1

$$\lambda = i$$

We have to solve the equation

$$A\mathbf{x} = i\mathbf{x}$$

Equivalently,

$$\begin{bmatrix} 0 & -1 \\ 1 & 0 \end{bmatrix} \begin{bmatrix} x \\ y \end{bmatrix} = \begin{bmatrix} ix \\ iy \end{bmatrix}$$

This matrix equation gives

$$-y = ix \quad \text{and} \quad x = iy$$

These equations are in fact the same, as can be observed by multiplying the first equation on both sides by i and simplifying, remembering that $i^2 = -1$.

Choosing $y = a$ implies that $x = ia$, and thus the eigenvector corresponding to the eigenvalue, i, is of the form

$$[ia \quad a]^T$$

Note that if you use the DERIVE function, EXACT_EIGENVECTOR, you will probably have the eigenvector in the form

$$[a \; - \; ia]^T$$

These are equivalent expressions for the same eigenvector.

Case 2

$$\lambda = - \, i$$

We have to solve the equation

$$A\mathbf{x} = - \, i\mathbf{x}$$

From which we obtain the equations

$$- \, y = - \, ix \quad \text{and} \quad x = - \, iy$$

Again these form the same equation, as can be observed by multiplying the first equation throughout by i. Hence the eigenvectors in this case have the form

$$[\text{-}ia \; a]^T$$

Check all these results using DERIVE before progressing to the second example of a rotation given below.

2.   In this case let us consider a rotation in $\mathfrak{R}^2$ through 180° about the origin, which is given by the matrix

$$\begin{bmatrix} -1 & 0 \\ 0 & -1 \end{bmatrix}$$

The characteristic equation is

$$(-1 - \lambda)^2 = 0$$

Again we have two solutions but they are equal, that is,

$$\lambda = -1 \text{ (twice)}$$

To find the eigenvectors we have to solve the matrix equation

$$A\mathbf{x} = -\mathbf{x}$$

This gives

$$\begin{bmatrix} -1 & 0 \\ 0 & -1 \end{bmatrix}\begin{bmatrix} x \\ y \end{bmatrix} = \begin{bmatrix} -x \\ -y \end{bmatrix}$$

From which we have the two equations

$$-x = -x \quad \text{and} \quad -y = -y$$

This means x can be anything and y can be anything.

How many independent vectors can we generate in this way in $\Re^2$?

For any basis in $\Re^2$ we need two independent vectors and we can certainly generate two from this problem. We could, of course, choose the standard basis $\{[1 \ 0], [0 \ 1]\}$ or any other in $\Re^2$. Hence there are two independent eigenvectors in this case and they can look like the standard basis vectors or any other two independent vectors in $\Re^2$.

Geometrically we see that although the x and y axes turn through 180°, they are in the same 'position' as when they started. This is true of any line through the origin, as that line is essentially a diameter of a circle.

This is the first example of an eigenvalue having two distinct eigenvectors associated with it.

7.2   EIGENSPACES

In this section we develop the idea of the preceding example of a 180° rotation where an eigenvalue gave us two independent eigenvectors. We bring together numerous ideas from earlier sections in this book and we would recommend you to read over this section a few times to establish in your own mind these important connections.

Recall the eigenvalue problem

$$(A - \lambda I)\mathbf{x} = \mathbf{0}$$

from which we defined the characteristic equation as

$$\text{DET}(A - \lambda I) = 0$$

We remarked that the system of equations was homogeneous but that we were looking for non-zero solutions for the eigenvectors, $\mathbf{x}$, associated with the eigenvalue, λ. In Section 6.6 we introduced the idea of the null space of a matrix. If we now compare these two ideas we find that one leads to the other. Solving the equation

$(A - \lambda I)\mathbf{x} = \mathbf{0}$

for non-zero eigenvectors $\mathbf{x}$ is identical to finding the non-zero vectors of the null space of the square matrix

$A - \lambda I$

As $A - \lambda I$ is an $n \times n$ matrix, we have seen in Section 6.6 that the null space of such a matrix is a subspace of $\Re^n$.

Thus given an eigenvalue, λ, we could have many eigenvectors, $\mathbf{x}$, satisfying the equation

$(A - \lambda I)\mathbf{x} = 0$

Definition

The null space of the square matrix $(A - \lambda I)$ is called *the eigenspace of* λ. Alternatively, we talk of

the eigenspace of A associated with the eigenvalue λ

In summary, what we have established is that the eigenvectors of A associated with the eigenvalue λ are the non-zero vectors in the null space of $(A - \lambda I)$. Furthermore, since the zero vector satisfies the equation $A\mathbf{x} = \lambda\mathbf{x}$ we have defined the eigenspace to be the null space of $(A - \lambda I)$ and this eigenspace is a vector subspace of $\Re^n$, where n is obtained from the size of the matrix A.

Consider again the 180° rotation in the preceding section where we found two independent vectors associated with the eigenvalue $\lambda = -1$. Equivalently we found that the null space of $A - \lambda I$, which in this case is $A + I$, was two-dimensional, and we established a basis for it. Thus the eigenspace associated with $\lambda = -1$ is two-dimensional and in fact was $\Re^2$ itself. It means in this case that every vector in $\Re^2$ was mapped to the zero vector by the matrix $(A - \lambda I)$.

It is now time to do some further problems to consolidate this very important section.

DERIVE ACTIVITY 7B

The first problem is taken from the DERIVE manual.

(A)   Consider the 3 × 3 matrix

$$\begin{bmatrix} 7 & -1 & -2 \\ -1 & 7 & -2 \\ -1 & -1 & 6 \end{bmatrix}$$

The problem is to find the eigenvalues and the associated eigenvectors using the functions, EIGENVALUES and EXACT_EIGENVECTOR.
    To do this you must first load VECTOR.MTH.
    Input the matrix in line 1 and ask for the eigenvalues in line 2. In line 3 you should obtain after simplifying line 2 the two values for eigenvalues, $w = 4$ and $w = 8$.
    (Remember w is the default label if you haven't declared a particular variable.)
    Now on line 4 input

EXACT_EIGENVECTOR (#1,4)

to obtain the corresponding eigenvector(s) for the eigenvalue, 4.
    On line 5 you should obtain something like

$$[x1 = @1 \quad x2 = @1 \quad x3 = @1]$$

This means that the eigenvector has the form

$$[a \ a \ a]^T$$

Now on line 6 input

EXACT_EIGENVECTOR (#1,8)

Simplifying you should have on line 7 something like

$$\left[x1 = @2 \quad x2 = @3 \quad x3 = -\frac{@2 + @3}{2} \right]$$

Interpreting this line, we see we have the third coordinate in terms of the first two coordinates. This means that we have some degrees of freedom.
How many?
If necessary at this stage refer back to Section 6.6
The answer is that there are two choices to be made, @2 and @3.
Let $x1 = 0$ and $x2 = 1$ so that $x3 = -1/2$.
Now let $x1 = 1$ and $x2 = 0$ so that $x3 = -1/2$.
We now have vectors of the form

$$[0 \ 1 \ -1/2] \quad \text{and} \quad [1 \ 0 \ -1/2]$$

These are independent by inspection and they span a two-dimensional space. Hence the eigenspace associated with the eigenvalue 8 is two-dimensional, whereas the eigenspace associated with the eigenvalue 4 is one-dimensional. This must mean that the eigenvalue 8 is a repeated root of the characteristic polynomial. Check this statement by inputting

CHARPOLY(#1)

and then factorizing or solving the cubic equation that you obtain.

Alternatively, for practice in evaluating determinants by hand using properties which reduce the entries, show *without* using DERIVE that

$$\begin{vmatrix} 7-\lambda & -1 & -2 \\ -1 & 7-\lambda & -2 \\ -1 & -1 & 6-\lambda \end{vmatrix} = (8-\lambda)^2(4-\lambda)$$

This is the characteristic polynomial for the matrix in line 1 and by putting it equal to zero we have the result obtained by DERIVE.

(B)   In this problem consider the 3×3 matrix

$$\begin{bmatrix} -5 & -5 & -9 \\ 8 & 9 & 18 \\ -2 & -3 & -7 \end{bmatrix}$$

Using DERIVE, or otherwise, find the eigenvalue(s) and the corresponding eigenvector(s) for this matrix.

Remember the characteristic polynomial must be of degree 3 and so we can expect three eigenvalues. Before looking at the solution below attempt the question on your own. Interpret your results.

Although the characteristic polynomial is of degree 3 you should find that when you factorize it DERIVE gives you

$$-(w + 1)^3$$

Thus there is just one eigenvalue, namely, –1. Alternatively,

$w = -1$ (three times)

If you then asked for the eigenvector by inputting

EXACT_EIGENVECTOR(#n,–1)

where n is the line number which has this matrix as its entry, you should find that DERIVE gives you something like

$$\left[x1 = @1 \quad x2 = -2@1 \quad x3 = \frac{2@1}{3} \right]$$

Thus the eigenvector associated with the eigenvalue –1 is of the form

[3 – 6 2]

There is one degree of freedom and thus the eigenspace is one-dimensional. Alternatively, the null space of the matrix ($A - \lambda I$) is one-dimensional and a basis for it is the vector given above.
Verify that ($A - \lambda I$) does indeed map this vector to the zero vector.
Also verify that A maps this vector to the negative of itself, that is,

$A\mathbf{x} = -\mathbf{x}$

Activity B has shown that the number of eigenvectors associated with the root $\lambda = -1$ (three times) does not equal 3, whereas in the first problem (A) we had a root occurring twice and we had two eigenvectors associated with it.   There is an important piece of theory coming from this observation.
First we will need a couple of definitions.   In equation theory we talk of repeated roots and the multiplicity of a root.
If a polynomial $f(x)$ of degree n has $(x - a)^m$ as a factor, then it can be written

$$f(x) = (x - a)^m g(x)$$

where $g(x)$ is a polynomial of degree $(n - m)$.

Definition

The algebraic multiplicity of the root $x = a$ of a polynomial $f(x) = 0$, where

$$f(x) = (x - a)^m g(x)$$

is m. We also say that $f(x)$ has a *repeated root*, namely, $x = a$.

As an example of how this idea is useful in describing eigenvalues, consider again the two characteristic equations of the two matrices in the above activity. In the first case we had, in factorized form,

$$(8 - \lambda)^2 (4 - \lambda) = 0$$

and in the second we had

$$(\lambda + 1)^3 = 0$$

In the first case we say that

the root $\lambda = 8$ has algebraic multiplicity 2

while

the root $\lambda = 4$ has algebraic multiplicity 1

In the second case we say that

the root $\lambda = -1$ has algebraic multiplicity 3

Note that there is another kind of multiplicity; hence the need to talk in terms of algebraic multiplicity in the above examples.

It follows from this idea that, counting multiplicities, every $n \times n$ matrix has exactly n eigenvalues in the set of complex numbers. We have seen this in all the examples in this chapter.

What was important above was whether the algebraic multiplicity of a particular eigenvalue had any correlation with the number of associated eigenvectors. The number of independent eigenvectors for a particular eigenvalue gave a basis for the eigenspace associated with that eigenvalue.

Definition

The geometric multiplicity of an eigenvalue λ is the dimension of the eigenspace associated with λ. Equivalently, the geometric multiplicity is the dimension of the null space of $(A - \lambda I)$.

Below we give a fact which relates the two multiplicities defined above.

Fact

The geometric multiplicity of an eigenvalue $\lambda \le$ the algebraic multiplicity of the same eigenvalue λ.

Referring back to DERIVE Activity 7B and the two examples given there, we can now state that in the first problem for the eigenvalue 8 the algebraic multiplicity was 2 and the geometric multiplicity was also 2.

In Problem 2 the algebraic multiplicity for the eigenvalue -1 was 3 but the geometric multiplicity was only 1.

7.3   SOME PROPERTIES OF EIGENVALUES

In this section we bring together some properties of eigenvalues which will be useful in computations.

1.   If λ is an eigenvalue of a linear transformation T and T^{-1} exists, then λ^{-1} is an eigenvalue of T^{-1}. Alternatively, we could replace T with the associated matrix A.

Proof

We are given that $T(\mathbf{v}) = \lambda\mathbf{v}$ (or $A\mathbf{x} = \lambda\mathbf{x}$). Now if T^{-1} exists we have

$$T^{-1}(T(\mathbf{v})) = T^{-1}(\lambda\mathbf{v})$$

Since T^{-1} is linear, $T^{-1}(\lambda\mathbf{v}) = \lambda T^{-1}(\mathbf{v})$.

Furthermore, $T^{-1}T = I$ (the identity transformation), and thus we have

$$T^{-1}(T(\mathbf{v})) = \mathbf{v} = T^{-1}(\lambda\mathbf{v}) = \lambda T^{-1}(\mathbf{v})$$

Finally, since λ is non-zero, λ^{-1} exists so

$$\mathbf{v} = \lambda T^{-1}(\mathbf{v})$$

becomes

$$\lambda^{-1}\mathbf{v} = \lambda^{-1}\lambda T^{-1}(\mathbf{v})$$

which simplifies to

$$\lambda^{-1}\mathbf{v} = T^{-1}(\mathbf{v})$$

This interprets as λ^{-1} is an eigenvalue of T^{-1}, as claimed.

2.   If A and B are $n \times n$ matrices, then AB and BA have the same eigenvalues.

Proof

Suppose that λ is an eigenvalue of the product AB. Hence,

$$AB\mathbf{x} = \lambda\mathbf{x}$$

where, as usual, $\mathbf{x}$ is a non-zero vector and λ is a non-zero eigenvalue.
If we now let $B\mathbf{x} = \mathbf{z}$, the equation becomes

$$AB\mathbf{x} = A\mathbf{z} = \lambda\mathbf{x}$$

Since $\mathbf{x} \neq 0$ and $\lambda \neq 0$ it follows that $\mathbf{z}$ cannot be zero.
Now $\mathbf{z}$ is an eigenvector of BA since the following chain of equalities is true:

$$BA\mathbf{z} = BAB\mathbf{x} = B\lambda\mathbf{x} = \lambda B\mathbf{x} = \lambda\mathbf{z}$$

Apart from this chain telling us that $\mathbf{z}$ is an eigenvector of BA we observe that the corresponding eigenvalue is λ.
Hence, having supposed that λ was an eigenvalue of AB we have arrived at the fact that λ is an eigenvalue of BA.
Similarly, if we begin with an eigenvalue for BA we can show that it will also be an eigenvalue of AB.
Thus AB and BA have the same eigenvalues.

3.   A matrix A and its transpose A^T have the same eigenvalues.

Proof

We prove that A and A^T have the same characteristic equation and hence the same eigenvalues. Start by considering the matrix

$$(A - \lambda I)^T$$

and use the properties of the transpose introduced earlier in the book.
We obtain the following equalities:

$$(A - \lambda I)^T = A^T - \lambda I^T = A^T - \lambda I$$

We also have the property of determinants that $DET(B) = DET(B^T)$ from earlier in the book. Hence,

$$DET(A - \lambda I) = DET((A - \lambda I)^T) = DET(A^T - \lambda I)$$

Therefore we have $DET(A - \lambda I) = DET(A^T - \lambda I) = 0$, which interprets as the characteristic equations of A and A^T are the same.

4.   Suppose $\mathbf{v}$ is an eigenvector of a linear transformation T with corresponding eigenvalue λ, then for $n > 0$, $\mathbf{v}$ is also an eigenvector of T^n with corresponding eigenvalue λ^n.

Proof

We are given that

$$T(\mathbf{v}) = \lambda\mathbf{v}$$

Now we use the composition of functions to evaluate

$$T(T(\mathbf{v})) = T(\lambda\mathbf{v}) = \lambda T(\mathbf{v})$$

The last equality is true because T is linear.
We write T composed with T as T^2 as usual, to obtain

$$T^2(\mathbf{v}) = \lambda T(\mathbf{v}) = \lambda(\lambda\mathbf{v}) = \lambda^2\mathbf{v}$$

Repeat this process to obtain

$$T(T^2(\mathbf{v})) = T(\lambda^2\mathbf{v})$$

which reduces to

$$T^3(\mathbf{v}) = \lambda^2 T(\mathbf{v}) = \lambda^2(\lambda\mathbf{v}) = \lambda^3\mathbf{v}$$

In n steps we will obtain

$$T^n(\mathbf{v}) = \lambda^n \mathbf{v}$$

as claimed.

5.   Similar matrices have the same eigenvalues.

Proof

Recall that two matrices A and B are similar if we can find a matrix P which has an inverse such that $A = P^{-1}BP$. Again what we prove is that similar matrices have the same characteristic equation.

Consider $\text{DET}(A - \lambda I) = \text{DET}(P^{-1}BP - \lambda I)$ from the above equality and let $D = \text{DET}(P^{-1}BP - \lambda I)$. Then

$D = \text{DET}(P^{-1}BP - P^{-1}\lambda IP)$	since λ is a scalar and $P^{-1}P = I$
$= \text{DET}(P^{-1}(B - \lambda I)P)$	by factorizing
$= \text{DET}(P^{-1})\text{DET}(B - \lambda I)\text{DET}(P)$	by property $\text{DET}(CD) = \text{DET}(C)\text{DET}(D)$
$= \text{DET}(P^{-1})\text{DET}(P)\text{DET}(B - \lambda I)$	since DET is a scalar and scalars commute
$= \text{DET}(P^{-1}P)\text{DET}(B - \lambda I)$	using the property above again
$= \text{DET}(I)\text{DET}(B - \lambda I)$	
$= 1 \cdot \text{DET}(B - \lambda I) = \text{DET}(B - \lambda I)$	

Hence,

$$\text{DET}(A - \lambda I) = \text{DET}(B - \lambda I)$$

from which it follows that the characteristic equations of A and B are the same. Thus A and B have the same eigenvalues, as claimed.

We will motivate the next property with an example.

6.   Below we find the characteristic equation for a general 2×2 matrix and observe some interesting results.

Reminder

The trace of a matrix is the sum of its diagonal elements (see Section 6.10).

Let

$$A = \begin{bmatrix} a & b \\ c & d \end{bmatrix}$$

Then

$$\mathrm{DET}(A - \lambda I) = \begin{vmatrix} a-\lambda & b \\ c & d-\lambda \end{vmatrix} = (a-\lambda)(d-\lambda) - bc$$

Collecting the terms together and equating $\mathrm{DET}(A - \lambda I)$ to zero we obtain the characteristic equation of A as

$$\lambda^2 - (a + d)\lambda + (ad - bc) = 0$$

Remark

Usually, the general quadratic equation is given as $ax^2 + bx + c = 0$. The roots x_1 and x_2 are given by the formulae

$$x_1 = \frac{-b+\sqrt{(b^2 - 4ac)}}{2a} \quad \text{and} \quad x_2 = \frac{-b-\sqrt{(b^2 - 4ac)}}{2a}$$

From these two formulae, two more can be obtained, namely,

$$x_1 + x_2 = -b/a \quad \text{and} \quad x_1 x_2 = c/a$$

Using these last two results with the roots λ_1 and λ_2 of our characteristic equation we obtain the following:

$$\lambda_1 + \lambda_2 = a + d$$

and

$$\lambda_1 \lambda_2 = ad - bc$$

Putting these results into words, we have

(a) the sum of the eigenvalues of a 2×2 matrix is the trace of A, and

(b)   the product of the two eigenvalues of a 2×2 matrix is DET(A).

These results are true for an $n \times n$ matrix and can be proved by using the theory of equations and considering the sum and product of the roots of a polynomial of the nth degree, as we indicate below.

Let $A = (a_{ij})$ be the general $n \times n$ matrix and then consider DET($A - \lambda I$):

$$DET(A - \lambda I) = \begin{vmatrix} a_{11} - \lambda & a_{12} & \cdots & a_{1n} \\ a_{21} & a_{22} - \lambda & \cdots & a_{2n} \\ \cdots & \cdots & \cdots & \cdots \\ a_{n1} & a_{n2} & \cdots & a_{nn} - \lambda \end{vmatrix}$$

Because an $n \times n$ matrix has n eigenvalues, if we count multiplicities, we can write

$$DET(A - \lambda I) = (\lambda_1 - \lambda)(\lambda_2 - \lambda) \cdots (\lambda_{n-1} - \lambda)(\lambda_n - \lambda)$$

To obtain the result (b) above we simply put $\lambda = 0$, so that

$$DET(A) = \lambda_1 \lambda_2 \cdots \lambda_{n-1} \lambda_n$$

that is, the determinant of a square matrix is equal to the product of its eigenvalues.

To obtain the result in (a) above we need to write DET($A - \lambda I$) in yet another form.

Consider the full determinant given above. If we start to expand it by the first row we would obtain the term

$$(a_{11} - \lambda)(a_{22} - \lambda) \cdots (a_{n-1n-1} - \lambda)(a_{nn} - \lambda)$$

together with other terms.

The other terms will all be polynomials in λ with degree $\leq (n - 2)$ since, for example, if we consider the cofactor of a_{12} we will have deleted the first row and the second column to obtain an $(n - 1) \times (n - 1)$ determinant.

In this process we will have deleted two terms in λ, namely, $(a_{11} - \lambda)$ and $(a_{22} - \lambda)$, and thus we cannot have any terms in λ^n or λ^{n-1} in the cofactor of a_{12}. Similarly with the rest of the elements in the first row. Hence we can write

$$\text{DET}(A - \lambda I) = (a_{11} - \lambda)(a_{22} - \lambda) \dots (a_{n-1n-1} - \lambda)(a_{nn} - \lambda)$$
$$+ \text{ a polynomial in } \lambda \text{ of degree} \le (n - 2)$$
$$= (-1)^n \lambda^n + (-1)^{n-1} \lambda^{n-1}(a_{11} + a_{22} + \dots + a_{nn})$$
$$+ \text{ a polynomial in } \lambda \text{ of degree} \le (n - 2)$$

From the quadratic case, we generalize, and have the following formula. The sum of the roots of a polynomial of degree n with variable x is

$$\frac{(-1)\text{coefficient of } x^{n-1}}{\text{coefficient of } x^n}$$

In the case of the quadratic equation the coefficient of x^{n-1} ($= x$) was given as b, and the coefficient of x^n ($= x^2$) was given as a; hence we arrive at the formula $-b/a$. In the expression for $\text{DET}(A - \lambda I)$ given above we see that this formula produces the following:

$$\frac{-(-1)^{n-1}(a_{11} + a_{22} + \dots + a_{nn})}{(-1)^n}$$

which simplifies to

$$a_{11} + a_{22} + \dots + a_{nn}$$

which in turn is the trace of the matrix A.

Hence the sum of the roots of the characteristic equation of a matrix A, $\lambda_1 + \lambda_2 + \dots + \lambda_{n-1} + \lambda_n$, is indeed the trace of the matrix A.

We conclude this section by checking these results on matrices from exercises given earlier in this book.

For the matrix

$$\begin{bmatrix} 0 & -1 \\ 1 & 0 \end{bmatrix}$$

We found the eigenvalues to be the complex conjugates, i and $-i$. By inspection, the trace of the matrix is zero and its determinant is 1. Thus the sum of the eigenvalues, zero, is the trace of the matrix. Furthermore, the product of the eigenvalues is $(i)(-i)$ $= -i^2 = +1$, since $i^2 = -1$, which agrees with the value of the determinant of the matrix.

Consider the matrix

$$\begin{bmatrix} -1 & 0 \\ 0 & -1 \end{bmatrix}$$

By inspection its trace is -2, while its determinant is 1. Earlier we found that the eigenvalues were -1 and -1. (Remember, multiplicities have to be counted.) Thus the sum is -2 and the product is $+1$, which again checks with our claims.

Now consider the first problem (A) in DERIVE Activity 7B, where we had the 3×3 matrix

$$\begin{bmatrix} 7 & -1 & -2 \\ -1 & 7 & -2 \\ -1 & -1 & 6 \end{bmatrix}$$

Again by inspection the trace is $(7 + 7 + 6) = 20$ and the determinant using DERIVE, or by hand, is found to be 256. The eigenvalues for this matrix were, 4, 8, and 8. Their sum is 20 and their product is 256, which again checks with our claims.

Finally, consider the second problem in DERIVE Activity 7B, where we had the matrix,

$$\begin{bmatrix} -5 & -5 & -9 \\ 8 & 9 & 18 \\ -2 & -3 & -7 \end{bmatrix}$$

By inspection the trace is $(-5 + 9 - 7) = -3$ and by hand or using DERIVE the determinant is -1. In the problem we had an eigenvalue of -1 with multiplicity 3, hence the sum of the eigenvalues is -3 and the product is -1, which again agree with our claims.

In the next chapter we will be looking again at eigenvalues and eigenvectors of special matrices.

Exercise 7C

Find the eigenvalues of the following matrices and show that their sum equals the trace and their product equals the value of the determinant:

$$\begin{bmatrix} 1 & 3 \\ 0 & 2 \end{bmatrix} \qquad \begin{bmatrix} 1 & 1 & 3 \\ -1 & 2 & 2 \\ 4 & 0 & -1 \end{bmatrix} \qquad \begin{bmatrix} 1 & 0 & 1 & 1 \\ 0 & 2 & 3 & -1 \\ 4 & -1 & 2 & 3 \\ -2 & 0 & 4 & 1 \end{bmatrix}$$

7.4   DIAGONALIZATION REVISITED

At the end of the previous chapter, in Section 6.10, we introduced the idea of similar matrices and when one matrix was similar to a diagonal matrix. In that section we asked two questions. We answered the second one by stating that not all $n \times n$ matrices are diagonalizable. We now answer the first question. (Recall Section 4.1 for the notation for a diagonal matrix.)

What we want to investigate is: Given an $n \times n$ matrix how can we tell whether it is diagonalizable or not? The answer, surprisingly or not, depends on the eigenvalues and eigenvectors of the given matrix. The first main result is stated below.

First Main Result

Let A be an $n \times n$ matrix with n linearly independent eigenvectors:

$$\mathbf{v}_1, \mathbf{v}_2, \ldots, \mathbf{v}_n$$

Now form a new matrix, B, with these eigenvectors as columns. Then B has an inverse and, furthermore, $B^{-1}AB$ is the matrix, D, given below:

$$\text{diag}(\lambda_1, \lambda_2, \ldots, \lambda_n)$$

where $A\mathbf{v}_i = \lambda_i \mathbf{v}_i$ for $1 \le i \le n$.

Proof

Since the $\mathbf{v}_i$s are linearly independent, B will have columns which are linearly independent and thus the column rank of B is n. Hence the row rank of B is also n. Thus it follows that $\text{DET}(B) \ne 0$ and thus B must have an inverse, B^{-1}.

Now we consider the matrix products AB and BD and show that they are equal, from which the result will follow.

First consider the product AB:

$$\begin{bmatrix} a_{11} & a_{12} & a_{13} & \cdots & a_{1n} \\ a_{21} & a_{22} & a_{23} & \cdots & a_{2n} \\ \cdots & \cdots & \cdots & \cdots & \cdots \\ a_{nl} & a_{n2} & a_{n3} & \cdots & a_{nn} \end{bmatrix} \begin{bmatrix} \mathbf{v}_1 & \mathbf{v}_2 & \cdots & \mathbf{v}_n \end{bmatrix}$$

Work out the first column of this product, assuming that $\mathbf{v}_1 = (x_1, x_2, ..., x_n)$. Pencil and paper would be useful at this point. You should be able to see that the first column of AB will simply be $A\mathbf{v}_1$.

But $A\mathbf{v}_1 = \lambda_1 \mathbf{v}_1$ so the product AB looks like the matrix

$$\begin{bmatrix} \lambda_1 \mathbf{v}_1 & \lambda_2 \mathbf{v}_2 & \cdots & \lambda_n \mathbf{v}_n \end{bmatrix}$$

This is an $n \times n$ matrix with columns $\lambda_i \mathbf{v}_i$.

Now consider the product BD:

$$\begin{bmatrix} \mathbf{v}_1 & \mathbf{v}_2 & \cdots & \mathbf{v}_n \end{bmatrix} \text{diag}(\lambda_1, \lambda_2, ..., \lambda_n)$$

Again we recommend using pencil and paper to write out this product more fully using the vector $\mathbf{v}_1$ as above and working out the first column of the product BD only.

You should obtain the result

$$\begin{bmatrix} \lambda_1 \mathbf{v}_1 & \lambda_2 \mathbf{v}_2 & \cdots & \lambda_n \mathbf{v}_n \end{bmatrix}$$

Hence,

$$AB = BD \quad \text{or, equivalently,} \quad B^{-1}AB = D$$

as claimed.

An alternative way of looking at this first main result is to say that any $n \times n$ matrix A, with n linearly independent eigenvectors, is diagonalizable. In the literature there are various reinterpretations or restatements of our first main result, all of which are equivalent. We have certainly given an answer to the question posed at the beginning of this section. We might now ask: do the eigenvalues tell us anything about whether or not a matrix is diagonalizable? The answer to this is given in our second main result below, but we have to be cautious when we talk in terms of eigenvalues, as we will show by examples later in this section.

Second Main Result

If v_1, v_2, ..., v_k are eigenvectors of an $n \times n$ matrix A and the associated eigenvalues λ_1, λ_2, ..., λ_k are distinct, that is, no multiplicities, then the v_is are linearly independent. It follows from this result that when $k = n$, k's maximum value, if A has n real distinct eigenvalues then A is diagonalizable. This is because if A has n real distinct eigenvalues then by our second main result A has n linearly independent eigenvectors, and by our first main result such a matrix is diagonalizable.

Before we proceed with caution, we give a proof of our second main result. You may like to revise Sections 3.8, 6.4 and 6.5 before looking at our proof.

Proof

Remark by way of revision. A set of k n-vectors will span a certain vector space V. From this span we can always find a minimal one which would then be a basis for V. The vectors forming this basis would therefore be linearly independent.

Now consider the k eigenvectors in our second main result. If we assume that they are not linearly independent, then we want to obtain a contradiction at the end of our argument. Assume, then, that our set is not linearly independent. We can therefore take a subset of the eigenvectors

$$\{v_1, v_2, ..., v_m\}$$

which is linearly independent, that is, a minimal span.

We can also write down a set of vectors which will be dependent, namely,

$$\{v_1, v_2, ..., v_m, v_{m+1}\}$$

where $1 \le m < m + 1 \le k$.

Consider, first, the dependent set. Because the vectors are dependent there exist scalars $a_1, a_2, ..., a_m, a_{m+1}$ which are not all zero, such that,

$$a_1 v_1 + a_2 v_2 + ... + a_m v_m + a_{m+1} v_{m+1} = 0 \tag{7.1}$$

Now multiply (7.1) throughout by the matrix A to obtain the equation

$$a_1 A v_1 + a_2 A v_2 + ... + a_m A v_m + a_{m+1} A v_{m+1} = A0 = 0$$

Since $A v_i = \lambda_i v_i$ for $1 \le i \le k$, we can rewrite this equation as

$$a_1\lambda_1\mathbf{v}_1 + a_2\lambda_2\mathbf{v}_2 + ... + a_m\lambda_m\mathbf{v}_m + a_{m+1}\lambda_{m+1}\mathbf{v}_{m+1} = \mathbf{0}$$

Now we return to Equation (7.1) and multiply throughout by λ_{m+1} in order to make the final terms in the equation above, and in the equation obtained below, the same:

$$a_1\lambda_{m+1}\mathbf{v}_1 + a_2\lambda_{m+1}\mathbf{v}_2 + ... + a_m\lambda_{m+1}\mathbf{v}_m + a_{m+1}\lambda_{m+1}\mathbf{v}_{m+1} = \mathbf{0}$$

Subtracting these last two equations will give us the equation

$$a_1(\lambda_1 - \lambda_{m+1})\mathbf{v}_1 + a_2(\lambda_2 - \lambda_{m+1})\mathbf{v}_2 + ... + a_m(\lambda_m - \lambda_{m+1})\mathbf{v}_m = \mathbf{0}$$

Notice at this stage that we have a linear combination of vectors from the linearly independent set only. Therefore, we can conclude that the coefficients are all equal to zero, that is,

$$a_i(\lambda_i - \lambda_{m+1}) = 0 \quad \text{for} \quad 1 \le i \le m$$

From our second main result we know that the eigenvalues are all distinct; therefore the parenthesized term can never be zero.

We are forced to conclude that

$$a_i = 0 \quad \text{for} \quad 1 \le i \le m$$

If we feed this information back into Equation (7.1) we obtain

$$a_{m+1}\mathbf{v}_{m+1} = \mathbf{0}$$

Since $\mathbf{v}_{m+1}$ is an eigenvector it cannot be the zero vector and thus we must conclude that a_{m+1} is zero, but that contradicts the fact that the set

$$\{\mathbf{v}_1, \mathbf{v}_2, ..., \mathbf{v}_m, \mathbf{v}_{m+1}\}$$

is dependent; hence our k eigenvectors must be independent, as claimed.

Now Our Word of Caution

When we have eigenvalues with multiplicities our second main result does not give us much information. Can such a situation still give us diagonalizability? The answer is yes and no!

The important point is that if an $n \times n$ matrix has n distinct eigenvalues it must have n linearly independent eigenvectors, but if an $n \times n$ matrix has less than n distinct eigenvalues it may or may not be diagonalizable.

Example 7E

In this example we shall use the matrices that we have seen before and about which we know quite a lot.

1.   For our first example look back to the matrix in Section 7.1:

$$A = \begin{bmatrix} 1 & 3 \\ 5 & 3 \end{bmatrix}$$

In that section we found that this matrix has two eigenvalues, 6 and −2. This means in terms of this section that our matrix A has two real distinct eigenvalues. As A is a 2×2 matrix our second main result tells us that this matrix is diagonalizable.

By our first main result we can write down the matrix B and D and check that $AB = BD$ or, alternatively, $B^{-1}AB = D$.

In Section 7.1 we found the eigenvectors to be [3 5] and [1 −1], taking $a = 1$, for corresponding eigenvalues 6 and −2.

Hence,

$$B = \begin{bmatrix} 3 & 1 \\ 5 & -1 \end{bmatrix} \quad \text{and} \quad D = \begin{bmatrix} 6 & 0 \\ 0 & -2 \end{bmatrix}$$

Observe that B does in fact have an inverse and checking by hand or by using DERIVE:

$$AB = BD = \begin{bmatrix} 18 & -2 \\ 30 & 2 \end{bmatrix}$$

2.   For our second example take the matrix representing the transformation, 'reflection in the y axis'. Here A is given by

$$A = \begin{bmatrix} -1 & 0 \\ 0 & 1 \end{bmatrix}$$

We found that A has eigenvalues +1 and −1 so again it must be diagonalizable. The corresponding eigenvectors we found to be [0 1] and [1 0] respectively. Hence we have for B and D respectively the following matrices:

$$B = \begin{bmatrix} 0 & 1 \\ 1 & 0 \end{bmatrix} \qquad D = \begin{bmatrix} 1 & 0 \\ 0 & -1 \end{bmatrix}$$

Again it is clear that B has an inverse and that

$$AB = BD = \begin{bmatrix} 0 & -1 \\ 1 & 0 \end{bmatrix}$$

3.   For our third example take the matrix representing a rotation in $\Re^2$ through 180° about the origin given by

$$A = \begin{bmatrix} -1 & 0 \\ 0 & -1 \end{bmatrix}$$

In this example we found that we had an eigenvalue of −1 with multiplicity 2. Furthermore, we found that although there was only one distinct eigenvalue we could produce two independent eigenvectors. Putting this another way, we found that the eigenspace was two-dimensional.

Now because A has two independent eigenvectors we know from our first main result that A is diagonalizable. Here then is the warning. Eigenvalues can give us some information but not all.

To finish this example we can write down B and D from the information we found in Section 7.1. They are, respectively,

$$\begin{bmatrix} 1 & 0 \\ 0 & 1 \end{bmatrix} \quad \text{and} \quad \begin{bmatrix} -1 & 0 \\ 0 & -1 \end{bmatrix}$$

As B turns out to be the identity it is immediate that it has an inverse and that $AB = BD$ since $A = D$ anyway!

4.   Our final example is from Section 7.2, DERIVE Activity 7B, and Problem (A). The matrix A is

$$A = \begin{bmatrix} 7 & -1 & -2 \\ -1 & 7 & -2 \\ -1 & -1 & 6 \end{bmatrix}$$

We found that A had eigenvalues 4 and 8, with 8 having multiplicity 2. Its eigenspace was found to be two-dimensional.

Hence once again we are on safe ground as we have three independent eigenvectors and thus can conclude that the matrix A above can be diagonalized.

If we make the eigenvectors have integer entries we can write down for B the matrix

$$B = \begin{bmatrix} 1 & 0 & 2 \\ 1 & 2 & 0 \\ 1 & -1 & -1 \end{bmatrix}$$

The matrix for D is

$$D = \begin{bmatrix} 4 & 0 & 0 \\ 0 & 8 & 0 \\ 0 & 0 & 8 \end{bmatrix}$$

Using DERIVE or by hand it easy to show that

$$AB = BD = \begin{bmatrix} 4 & 0 & 16 \\ 4 & 16 & 0 \\ 4 & -8 & -8 \end{bmatrix}$$

Check, finally, that B does have an inverse!

Consolidation Exercise

1. Show that the characteristic equation of the matrix A given below is the cubic $\lambda^3 - 3\lambda^2 - 9\lambda - 5 = 0$:

$$A = \begin{bmatrix} 1 & 2 & 2 \\ 2 & 1 & 2 \\ 2 & 2 & 1 \end{bmatrix}$$

Verify that the Cayley–Hamilton theorem holds for the matrix A. Given that the eigenvalues of A are all integers, calculate them. Hence show that A satisfies not just the cubic equation but a certain polynomial of the second degree.

(Observe that A is symmetric and the eigenvalues are real. In Chapter 8 we will be reminded of this fact.)

Check your result using DERIVE.

2.   Let $T: \Re^2 \to \Re^2$ be given by $T(x,y) = (3x + 3y, x + 5y)$. Show that this defines a linear transformation.

Find all the eigenvalues of T and compute the corresponding eigenvectors. Describe the corresponding eigenspaces.

3.   Repeat Problem 2 above but with T defined as $T(x,y) = (y,x)$.

4.   Repeat Problem 2 but with T defined as $T(x,y) = (y,-x)$.

5.   Consider each of the matrices given below and decide whether or not they are diagonalizable. Call the matrix given, A.

If A is diagonalizable find an invertible matrix P and a diagonal matrix D such that $P^{-1}AP = D$.

Some parts are easily done without DERIVE.

We recommend that you do them all by hand before checking your results using DERIVE:

(i) $\begin{bmatrix} 2 & -4 \\ 1 & -2 \end{bmatrix}$ (ii) $\begin{bmatrix} 1 & 4 \\ 1 & -2 \end{bmatrix}$ (iii) $\begin{bmatrix} 1 & 0 & 0 \\ 0 & 1 & 2 \\ 0 & 0 & 2 \end{bmatrix}$ (iv) $\begin{bmatrix} 1 & 0 & 0 \\ 1 & 1 & 0 \\ 0 & 1 & 1 \end{bmatrix}$

8

Orthogonal Matrices

8.1   SOME REMINDERS AND DEFINITIONS

You will need to recall the following ideas from Chapter 3 before proceeding. Vectors, norm or length of a vector, scalar or dot product and, in particular, Section 3.5 on the orthogonality of vectors. So what do we mean by an orthogonal matrix?

Is it one where the rows (or columns) thought of as vectors have the property that any two rows (or columns) have for their scalar product the value of zero? Recall that two vectors whose scalar product is zero are at right angles to each other, or are orthogonal.

Definition

An $n \times n$ matrix A, with real entries, is said to be *orthogonal* if

$$A^T = A^{-1}$$

In other words, a square matrix is orthogonal if its transpose and its inverse are the same.

The above definition can be restated in the following way: a matrix A is orthogonal if

$$A^T A = A A^T = I$$

This follows immediately from the definition of an inverse.

If the entries of the matrix are complex numbers, then we use different names for certain matrices. In Chapter 1, Section 1.5, we defined the idea of a symmetric matrix and assumed that it had real entries. Consider the matrix with complex entries given below:

$$\begin{bmatrix} 2 & 2+i & 5 \\ 2-i & 7 & 5-3i \\ 5 & 5+3i & 8 \end{bmatrix}$$

In some sense this matrix is symmetric!

In complex numbers we have the idea of a complex conjugate. Recall that if $z = a + ib$ is a complex number then $\bar{z} = a - ib$ is its conjugate, and vice versa. We speak of 'z bar' or z-conjugate when we read $\bar{z}$.

We observe in the matrix above that a_{ij} and a_{ji} are either real numbers or complex conjugates of each other. Consider a_{12} and a_{21}, for example. They are $2 + i$ and $2 - i$, respectively, whereas a_{13} and a_{31} are both purely real and equal to 5.

Notation

We will denote by $A*$ the matrix obtained from A by replacing complex number entries by their conjugates and then transposing the resulting matrix. In other words, if $A = (a_{ij})$, then $A* = (\overline{a_{ji}})$:

$$\text{If } A = \begin{bmatrix} 1 & 2+i \\ 3 & 1-3i \end{bmatrix} \text{ then } A* = \begin{bmatrix} 1 & 3 \\ 2-i & 1+3i \end{bmatrix}$$

In the literature $A*$ is often referred to as the *conjugate transpose* of A.

Definition

A matrix A with complex entries is said to be *Hermitian* if $A* = A$. This is the counterpart to the symmetric matrix where real entries are present. The counterpart to the orthogonal matrix where complex number entries occur is given below.

Definition

An $n \times n$ matrix A with complex entries is said to be *unitary* if

$$A* = A^{-1}$$

In other words, a matrix is unitary if its conjugate transpose is equal to its inverse. Again, the alternative definition is: a matrix A is unitary if

$$A*A = AA* = I$$

Summary

A matrix is symmetric or Hermitian if $A^T = A$ or $A* = A$ respectively.
A matrix is orthogonal or unitary if $A^T = A^{-1}$ or $A* = A^{-1}$ respectively.

Consider again the 3×3 matrix

$$\begin{bmatrix} 2 & 2+i & 5 \\ 2-i & 7 & 5-3i \\ 5 & 5+3i & 8 \end{bmatrix}$$

This is Hermitian because the 'symmetric elements' are conjugates of each other.

Example 8A

In this exercise we will answer the question posed at the start of this section. In other words,

Are the rows and columns of an orthogonal matrix special?

Solution

Consider the general 2×2 matrix

$$\begin{bmatrix} a & b \\ c & d \end{bmatrix}$$

Let us assume that it is orthogonal.

Note that we could have assumed that it was unitary, which covers the orthogonal case when complex numbers are purely real, but for simplicity we assume that the entries a, b, c and d are real. Since the matrix is orthogonal we have the following true equation:

$$\begin{bmatrix} a & c \\ d & d \end{bmatrix} = \frac{1}{ad-bc} \begin{bmatrix} d & -b \\ -c & a \end{bmatrix}$$

Alternatively, we have the equation $A^T A = AA^T = I$, which gives, for $A^T A = I$,

$$\begin{bmatrix} a & c \\ b & d \end{bmatrix} \begin{bmatrix} a & b \\ c & d \end{bmatrix} = \begin{bmatrix} 1 & 0 \\ 0 & 1 \end{bmatrix}$$

Expanded out we have the following four equations:

$$a^2 + c^2 = 1$$

$$ab + cd = 0$$

$$ba + dc = 0$$

$$b^2 + d^2 = 1$$

Note that the second and third equations are the same. Now consider the two columns of matrix A as vectors:

$$[a \ c]^T \quad \text{and} \quad [b \ d]^T$$

The scalar product of these vectors is, from Chapter 3,

$$ab + cd$$

but from the above equations we see that this is equal to zero and thus the two vectors are orthogonal. This means that the two columns thought of as vectors are orthogonal to each other.

What do the other two equations tell us about the vectors

$$[a \ c]^T \quad \text{and} \quad [b \ d]^T?$$

The norm or length of the vector $[a \ c]^T$ is $\sqrt{(a^2 + c^2)}$ using Chapter 3 again and similarly, the norm or length of the vector $[b \ d]^T$ is $\sqrt{(b^2 + d^2)}$. This means that the lengths of these two vectors are one, from the equations found above.

We now introduce another new word by way of a definition.

Definition

A set of vectors $\{v_i\}$ is said to be *orthonormal* if for every pair of vectors in the set, their scalar product is zero *and* the length of each vector v_i, is equal to one. In other words, every pair of vectors is at right angles to every other or is orthogonal, and each vector has unit length.

In symbols we have $v_i.v_j = 0$ and $|v_i| = 1$ for all i and j. The pair of column vectors

$$[a \ c]^T \quad \text{and} \quad [b \ d]^T$$

has therefore been proved to be an orthonormal set. Similarly, if you begin with the equation $AA^T = I$ you should obtain the equations

$$a^2 + b^2 = 1 = c^2 + d^2 \quad \text{and} \quad ac + bd = 0$$

These can be interpreted again by considering the rows of A as vectors:

$$[a \quad b] \quad \text{and} \quad [c \quad d]$$

The equations now mean that these vectors are of unit length and that their scalar product is zero. Hence the rows of A considered as vectors form an orthonormal set.

Summary

It follows from this exercise that if a 2×2 matrix is orthogonal, then its rows form an orthonormal set and its columns form another orthonormal set. The above result holds true for $n \times n$ orthogonal matrices and answers the question posed at the start of this section.

   The same result is true if we replace orthogonal by unitary, and we leave the reader who is familiar with complex numbers to follow Example 8A through again, with the original 2×2 general matrix being assumed to be unitary.

Exercise 8A

1.   Show that for a 2×2 unitary matrix the rows and columns form orthonormal sets.
2.   Show that if A is an orthogonal matrix and B is a symmetrix matrix, then $A^{-1}BA$ is a symmetric matrix.
3.   Find the values of α so that the matrix C is unitary:

$$C = \alpha \begin{bmatrix} 1 & 1+i \\ 1-i & -1 \end{bmatrix}$$

4.   If a matrix H is Hermitian show that the eigenvalues of H are real.

DERIVE ACTIVITY 8A

(1)   Consider again the general 2×2 matrix used in Example 8A. Assuming that
it was orthogonal we arrived at the following true equation:

$$\begin{bmatrix} a & c \\ b & d \end{bmatrix} = \frac{1}{ad - bc} \begin{bmatrix} d & -b \\ -c & a \end{bmatrix}$$

In Example 8A we used the alternative form because it was easier to handle.
We will now use DERIVE to handle the above equation.

     Author the general matrix on line 1.

     Now obtain the transpose and the inverse of this general matrix. Now
put them equal on your next line and **Simplify**.

     Your screen should look like that shown in Figure 8.1.

#6: $\begin{bmatrix} a & c \\ b & d \end{bmatrix} = \begin{bmatrix} \dfrac{d}{a \cdot d - b \cdot c} & \dfrac{c}{b \cdot c - a \cdot d} \\ \dfrac{b}{b \cdot c - a \cdot d} & \dfrac{a}{a \cdot d - b \cdot c} \end{bmatrix}$

#7: $\begin{bmatrix} a = \dfrac{d}{a \cdot d - b \cdot c} & c = \dfrac{c}{b \cdot c - a \cdot d} \\ b = \dfrac{b}{b \cdot c - a \cdot d} & d = \dfrac{a}{a \cdot d - b \cdot c} \end{bmatrix}$

```
COMMAND: Author Build Calculus Declare Expand Factor Help Jump soLve Manage
         Options Plot Quit Remove Simplify Transfer Unremove moVe Window approX
Compute time: 0.1 seconds
Simp(#6)                                    Free:48%                Derive Algebra
```

Figure 8.1   Screen dump for an orthogonal matrix

The diagonal elements can be written in terms of DET(A):

     $a = d/\text{DET}(A)$ and $d = a/\text{DET}(A)$

Hence,

     $a = a/(\text{DET}(A))^2$

from which we conclude that

     $a = 0$ or $\text{DET}(A)^2 = 1$

In general $a \neq 0$, so $\text{DET}(A) = \pm 1$.

Similarly, if we consider the other two elements we have

$$c = \frac{b}{-\text{DET}(A)} \quad \text{and} \quad b = \frac{c}{-\text{DET}(A)}$$

Hence,

$$b = b/(\text{DET}(A))^2$$

On the assumption that $b \neq 0$, in general, we arrive at the same condition on DET(A). In theory we can obtain the same result, as follows.

Consider an $n \times n$ orthogonal matrix A, so that $AA^T = A^TA = I$. Now take determinants of these equations to obtain

$$\text{DET}(AA^T) = \text{DET}(A^TA) = \text{DET}(I)$$

Using the property of determinants seen in Chapter 2, namely,

$$\text{DET}(CD) = \text{DET}(C)\text{DET}(D)$$

we have the following:

$$\text{DET}(A)\text{DET}(A^T) = \text{DET}(A^T)\text{DET}(A) = 1$$

since the determinant of the identity matrix is 1.

Furthermore, from determinants we have that $\text{DET}(A) = \text{DET}(A^T)$ and thus we have

$$\text{DET}(A)^2 = 1$$

(B)   Calculate the determinant of the following matrices. If the determinant $= \pm 1$, then decide if the matrix is orthogonal.

$$\begin{bmatrix} 1 & -2 & -2 \\ -2 & 1 & -2 \\ -2 & -2 & 0 \end{bmatrix} \qquad \begin{bmatrix} 1 & 0 & 0 \\ 0 & 0.6 & 0.8 \\ 0 & -0.8 & 0.6 \end{bmatrix} \qquad \begin{bmatrix} 1 & 0 & 7 \\ -3 & 2 & 2 \\ -2 & 1 & -2 \end{bmatrix}$$

Fact

If A is an $n \times n$ *orthogonal* matrix, then DET(A) = ±1. This supports our DERIVE activity and is true for any orthogonal matrix. This fact also gives us a quick check as to whether a matrix A is possibly orthogonal. If DET(A) ≠ ±1, then A cannot be orthogonal. However, if DET(A) = ±1 we cannot conclude that A is necessarily orthogonal.

Consider the 2×2 matrix

$$\begin{bmatrix} 1 & 1 \\ 0 & 1 \end{bmatrix}$$

It has determinant +1 but it is not orthogonal since its transpose does not equal its inverse. Check this assertion for yourself.

We will now investigate the general 2×2 case and see what orthogonal matrices look like. We have already done much of the groundwork above.

Consider the general 2×2 matrix given in Example 8A and the equation given at the start of DERIVE Activity 8A, that is,

$$\begin{bmatrix} a & c \\ b & d \end{bmatrix} = \frac{1}{ad - bc} \begin{bmatrix} d & -b \\ -c & a \end{bmatrix}$$

Since DET(A) = ±1 we have two cases to consider.

Case 1

$$\text{DET}(A) = +1$$

Then from the above matrix equation we have just two equations, namely,

$$a = d \quad \text{and} \quad c = -b.$$

Hence, the matrix A has the form

$$\begin{bmatrix} a & b \\ -b & a \end{bmatrix}$$

with determinant $a^2 + b^2 = 1$.

Typically, the solution could be written

$$a = \cos(\theta) \quad \text{and} \quad b = \sin(\theta)$$

The final form of the general matrix would now be

$$\begin{bmatrix} \cos(\theta) & \sin(\theta) \\ -\sin(\theta) & \cos(\theta) \end{bmatrix}$$

We have already seen this matrix in Chapter 6, Section 6.7, as the general rotation matrix in $\Re^2$. These matrices are one type of orthogonal matrix.

We now consider the other type.

Case 2

$$DET(A) = -1$$

From the matrix equation we now have the two equations

$$a = -d \quad \text{and} \quad b = c$$

The general matrix now has the form

$$\begin{bmatrix} a & b \\ b & -a \end{bmatrix}$$

with determinant $-a^2 - b^2 = -1$, that is, $a^2 + b^2 = 1$.

Again, we could use $a = \cos(\theta)$ and $b = \sin(\theta)$ to give the general matrix

$$\begin{bmatrix} \cos(\theta) & \sin(\theta) \\ \sin(\theta) & -\cos(\theta) \end{bmatrix}$$

This in general represents a composition of a reflection matrix with a rotation matrix.

Hence any orthogonal 2×2 matrix is either a rotation or reflection (or both). Again refer to Chapter 6, Section 6.7, for examples of these types.

8.2   ORTHOGONAL BASES

In this section we are going to explore orthogonal vectors and matrices further. In Chapter 6, Section 6.4, we met the idea of a basis for a vector space as a set of linearly independent vectors.

Definition

An *orthogonal basis* is a basis in which the vectors are mutually orthogonal. This means that any two vectors in the basis are orthogonal.

Fact

If $\{v_i\}$ is a set of non-zero orthogonal vectors, then they are a linearly independent set. This means that they are an orthogonal basis for some vector space. We prove this fact below.

Proof

Consider the set $\{v_i\}$ of n, non-zero, orthogonal vectors. We write down a linear combination of them and equate it to zero. Refer to Chapter 6 to refresh your memory. Hence we have

$$\alpha_1 v_1 + \alpha_2 v_2 + \ldots + a_{n-1} v_{n-1} + \alpha_n v_n = 0$$

We have to prove that every $\alpha_i = 0$.
     Next, scalar product each side by v_i to obtain

$$\alpha_1 v_1 \cdot v_i + \alpha_2 v_2 \cdot v_i + \ldots + a_{n-1} v_{n-1} \cdot v_i + \alpha_n v_n \cdot v_i = 0 . v_i$$

This equation simplifies to

$$\alpha_1 0 + \alpha_2 0 + \ldots + \alpha_i v_i \cdot v_i + \ldots + \alpha_{n-1} 0 + \alpha_n 0 = 0$$

since v_i is orthogonal to every other vector in the set. This reduces to the equation

$$\alpha_i v_i \cdot v_i = 0$$

But v_i is non-zero so we must conclude that $\alpha_i = 0$, for all i, which proves the fact that we stated above.
     The standard basis for $\Re^n$:

$$[1 \quad 0 \quad 0 \quad \ldots \quad 0], [0 \quad 1 \quad 0 \quad \ldots \quad 0], \quad \ldots \quad ,[0 \quad 0 \quad 0 \quad \ldots \quad 0 \quad 1]$$

is an orthogonal basis since the vectors in it are mutually orthogonal. Furthermore, it is an orthonormal basis since each of the vectors is of unit length. It is perhaps the most famous orthonormal basis in linear algebra.

8.3   THE GRAM–SCHMIDT ORTHOGONALIZATION PROCESS

Although the title of this section is a little forbidding, the idea is a relatively straightforward one. The idea is that we are given a basis for a vector space on which a scalar or dot product is defined, and we wish to change that basis into an orthonormal one. If we are given a basis for $\mathfrak{R}^n$ it must consist of n independent vectors. Let such a basis be $\{v_1, v_2, ..., v_n\}$. We want to use these vectors to produce a set of vectors all of unit length, any two of which will be orthogonal, that is, an orthonormal basis. Let such a basis be labelled $\{u_1, u_2, ..., u_n\}$. The use of 'u' will remind us that we are producing unit vectors. From the preceding section we know that such a set will in fact be independent.

We start by producing u_1. Take the vector v_1. It is not likely to be a unit vector so we make it of length one by dividing it by its length, $|v_1|$, see Section 3.2. Hence we can put

$$u_1 = \frac{v_1}{|v_1|}$$

Next we want to use u_1, which we have just produced, together with v_2 from our original set, to produce u_2. We want u_2 to be orthogonal to u_1, that is, $u_1 \cdot u_2 = 0$.

We can picture the vectors v_1, u_2, and v_2 as a right angled triangle with v_2 as the hypotenuse, and both v_1 and v_2 starting from the same point. We can write v_1 as ku_1 and from the triangle we know that $k = |v_2| \cos(\theta)$, where θ is the angle between v_1 and v_2. See Figure 8.2.

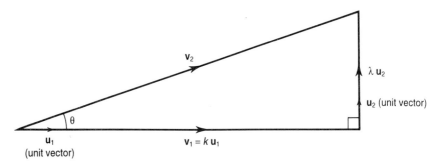

Figure 8.2   The vectors v_1, u_2 and v_2 form a right angled triangle

Furthermore, from the triangle we have the vector equation

$$v_1 + \lambda u_2 = v_2$$

where λ is some scalar yet to be obtained. Thus we have

$$\lambda\mathbf{u}_2 = \mathbf{v}_2 - \mathbf{v}_1$$

$$= \mathbf{v}_2 - k\mathbf{u}_1$$

$$= \mathbf{v}_2 - |\mathbf{v}_2|\cos(\theta)\mathbf{u}_1$$

Thus,

$$\lambda\mathbf{u}_2 = \mathbf{v}_2 - (\mathbf{u}_1 \cdot \mathbf{v}_2)\mathbf{u}_1$$

Recall that the scalar or dot product of two vectors, **a** and **b**, is given by

$$\mathbf{a} \cdot \mathbf{b} = |\mathbf{a}||\mathbf{b}|\cos(\alpha)$$

where α is the angle between the vectors **a** and **b**.

From the above equation we can see that λ is the magnitude of the vector

$$\mathbf{v}_2 - (\mathbf{u}_1 \cdot \mathbf{v}_2)\mathbf{u}_1$$

We can check directly that $\lambda\mathbf{u}_2$ and $\mathbf{u}_1$ are orthogonal by taking dot products, that is,

$$(\lambda\mathbf{u}_2 \cdot \mathbf{u}_1) \qquad = (\mathbf{v}_2 \cdot \mathbf{u}_1) - (\mathbf{u}_1 \cdot \mathbf{v}_2)(\mathbf{u}_1 \cdot \mathbf{u}_1) = 0$$

since $(\mathbf{u}_1 \cdot \mathbf{u}_1) = 1$ and $(\mathbf{u}_1 \cdot \mathbf{v}_2) = (\mathbf{v}_2 \cdot \mathbf{u}_1)$ in $\Re^n$. Thus we have constructed $\mathbf{u}_2$.

We continue in a similar way to produce $\mathbf{u}_3$ and this can still be visualized as we are now into $\Re^3$. To visualize where $\mathbf{u}_4$ would be is somewhat difficult!

We produce $\mathbf{u}_3$ from the vectors $\mathbf{u}_1$, $\mathbf{u}_2$ and $\mathbf{v}_3$. At this stage draw a picture with $\mathbf{u}_1$, $\mathbf{u}_2$ and $\mathbf{u}_3$ as the traditional x, y and z axes. The vector $\mathbf{v}_3$ will start at the origin and point into space. Join appropriate lines to make a rectangular-based pyramid. See Figure 8.3, from which you should be able to obtain the vector equation

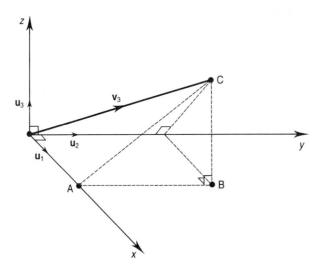

Figure 8.3   Visualization of $\mathbf{u}_1$, $\mathbf{u}_2$, $\mathbf{u}_3$ and $\mathbf{v}_3$

$$\mathbf{v}_3 = (\mathbf{u}_2 \cdot \mathbf{v}_3)\mathbf{u}_2 + (\mathbf{u}_1 \cdot \mathbf{v}_3)\mathbf{u}_1 + \mu\mathbf{u}_3$$

from which we can write

$$\mu\mathbf{u}_3 = \mathbf{v}_3 - (\mathbf{u}_1 \cdot \mathbf{v}_3)\mathbf{u}_1 - (\mathbf{u}_2 \cdot \mathbf{v}_3)\mathbf{u}_2$$

Thus we have constructed $\mathbf{u}_3$, with μ being the magnitude of the right-hand side. Furthermore, by taking dot products of both sides with $\mathbf{u}_1$ and $\mathbf{u}_2$ separately you can verify that $\mathbf{u}_1$ is orthogonal to $\mathbf{u}_3$ and $\mathbf{u}_2$ is orthogonal to $\mathbf{u}_3$. Consider the proof of the second of these.

Taking dot products of $\mu\mathbf{u}_3$ with $\mathbf{u}_2$ we obtain

$$(\mu\mathbf{u}_3 \cdot \mathbf{u}_2) = (\mathbf{v}_3 \cdot \mathbf{u}_2) - (\mathbf{u}_1 \cdot \mathbf{v}_3)(\mathbf{u}_1 \cdot \mathbf{u}_2) - (\mathbf{u}_2 \cdot \mathbf{v}_3)(\mathbf{u}_2 \cdot \mathbf{u}_2)$$

Knowing that $(\mathbf{u}_1 \cdot \mathbf{u}_2) = 0$, $(\mathbf{u}_2 \cdot \mathbf{u}_2) = 1$ and $(\mathbf{v}_3 \cdot \mathbf{u}_2) = (\mathbf{u}_2 \cdot \mathbf{v}_3)$ we see that the right-hand side of this equation is in fact zero, as required. From consideration of the two expressions

$$\lambda\mathbf{u}_2 = \mathbf{v}_2 - (\mathbf{u}_1 \cdot \mathbf{v}_2)\mathbf{u}_1$$

and

$$\mu\mathbf{u}_3 = \mathbf{v}_3 - (\mathbf{u}_1 \cdot \mathbf{v}_3)\mathbf{u}_1 - (\mathbf{u}_2 \cdot \mathbf{v}_3)\mathbf{u}_2$$

We can continue and produce u_4 from u_1, u_2, u_3 and v_4.

A similar equation to the ones given above is given by

$$vu_4 = v_4 - (u_1 \cdot v_4)u_1 - (u_2 \cdot v_4)u_2 - (u_3 \cdot v_4)u_3$$

This produces u_4 and v is the magnitude of the right-hand side. Furthermore, it is a simple matter of checking that this vector is indeed orthogonal to u_1, u_2 and u_3, in a similar manner to the examples given above.

The General Situation

Let V be a vector space with a well-defined scalar product $(a \cdot b)$. In general, if $\{u_1, u_2, \ldots u_r\}$ is a set of orthonormal vectors, then for any $v \in V$, the vector w given by

$$w = v - (u_1 \cdot v)u_1 - (u_2 \cdot v)u_2 \ldots - (u_r \cdot v)u_r$$

is orthogonal to each of the u_is for $1 \le i \le r$.

Example 8B

Given a basis for $\mathfrak{R}^3$ as $\{(1, 1, 1), (0, 1, 1), (0, 0, 1)\}$, transform this basis into an orthonormal basis using the Gram–Schmidt process.

Solution

First take $v_1 = (1, 1, 1)$. This vector is of length, $\sqrt{3}$. Hence we define

$$u_1 = \frac{v_1}{|v_1|} = \frac{1}{\sqrt{3}}(1, 1, 1)$$

So

$$u_1 = \left(1/\sqrt{3}, \ 1/\sqrt{3}, \ 1/\sqrt{3}\right)$$

Now consider

$$\lambda u_2 = v_2 - (u_1 \cdot v_2)u_1$$

Recall that

$$(\mathbf{u}_1 \cdot \mathbf{v}_2) = \left(1/\sqrt{3},\ 1/\sqrt{3},\ 1/\sqrt{3}\right)(0,\ ,1,\ 1)^T = 2/\sqrt{3}$$

Substitute the values we know and we obtain the equation

$$\lambda\mathbf{u}_2 = (0,\ 1,\ 1) - \frac{2}{\sqrt{3}}\left(1/\sqrt{3},\ 1/\sqrt{3},\ 1/\sqrt{3}\right)$$

Therefore,

$$\lambda\mathbf{u}_2 = (-2/3,\ 1/3,\ 1/3).$$

This is not normalized, that is, it is not of unit length, but λ is the magnitude of the right-hand side and, by direct computation,

$$\lambda = \sqrt{(4/9 + 1/9 + 1/9)} = \sqrt{(6/9)} = \sqrt{(2/3)}$$

Hence we have $\mathbf{u}_2$ given as

$$\frac{1}{\sqrt{2/3}}(-2/3,\ 1/3,\ 1/3) = \left(-2/\sqrt{6},\ 1/\sqrt{6},\ 1/\sqrt{6}\right)$$

$$\mathbf{u}_2 = \left(-2/\sqrt{6},\ 1/\sqrt{6},\ 1/\sqrt{6}\right)$$

Finally, we obtain $\mathbf{u}_3$ from the equation

$$\mu\mathbf{u}_3 = \mathbf{v}_3 - (\mathbf{u}_1 \cdot \mathbf{v}_3)\mathbf{u}_1 - (\mathbf{u}_2 \cdot \mathbf{v}_3)\mathbf{u}_2$$

Again work out the scalar products first, then substitute all the known values into the right-hand side. You should obtain the following:

$$\mu\mathbf{u}_3 = (0,\ 0,\ 1) - 1/\sqrt{3}\left(1/\sqrt{3},\ 1/\sqrt{3},\ 1/\sqrt{3}\right) - 1/\sqrt{6}\left(-2/\sqrt{6},\ 1/\sqrt{6},\ 1/\sqrt{6}\right)$$

which simplifies to

$$\mu\mathbf{u}_3 = (0,\ -1/2,\ 1/2)$$

The length of the vector $(0,\ -1/2,\ 1/2)$ is

$$\sqrt{(0 + 1/4 + 1/4)} = \sqrt{(1/2)} = 1/\sqrt{2}$$

that is,

$$\mathbf{u}_3 = (0, -1/\sqrt{2}, 1/\sqrt{2})$$

Thus the required orthonormal basis is

$$\mathbf{u}_1 = (1/\sqrt{3}, 1/\sqrt{3}, 1/\sqrt{3}), \quad \mathbf{u}_2 = (-2/\sqrt{6}, 1/\sqrt{6}, 1/\sqrt{6}), \quad \mathbf{u}_3 = (0, -1/\sqrt{2}, 1/\sqrt{2})$$

Exercise 8B

1.   Check that $|\mathbf{u}_i| = 1$ for all i and that $(\mathbf{u}_i \cdot \mathbf{u}_j) = 0$ for all i and j.
2.   Repeat Example 8B for the basis

$$\{(1,0,0), (1,0,1), (1,1,0)\}$$

DERIVE ACTIVITY 8B

(A)   We will now answer the question posed in Example 8A above but using DERIVE to do the arithmetic. Input the vectors given in the first three lines, namely,

[1,1,1], [0,1,1], [0,0,1]

In line 4 find the length or norm of the vector [1,1,1] by **Author**

ABS(#1)

Simplify to see $\sqrt{3}$ in line 5.
Similarly, find the length of [0,1,1] in line 6 and simplify to $\sqrt{2}$ in line 7.
Finally, find the length of [0,0,1] in line 8 and simplify to obtain 1 in line 9.
In line 10 we normalize the vector [1,1,1] by dividing this vector by its length. A command such as #1/#5 should get the required result.
Simplify in line 11 to see the vector $\mathbf{u}_1$ obtained previously.
So far you might think it quicker to do the work by hand as the numbers are not too difficult to manipulate. At this stage DERIVE comes into its own by giving us $\lambda\mathbf{u}_2$ quickly.
On line 12 **Author** $\mathbf{v}_2 - (\mathbf{u}_1 \cdot \mathbf{v}_2)\mathbf{u}_1$ in the form

$$\#2 - (\#11.\#2)(\#11)$$

Simplify reveals $\lambda \mathbf{u}_2$ as in the exercise. On lines 14 and 15 find the length of this vector, $\sqrt{6}/3$.

Now to obtain $\mathbf{u}_2$ divide line 13 by line 15 and simplify. On line 17 we have $\mathbf{u}_2$ immediately.

On line 18 **Author**, $\mathbf{v}_3 - (\mathbf{u}_1.\mathbf{v}_3)\mathbf{u}_1 - (\mathbf{u}_2.\mathbf{v}_3)\mathbf{u}_2$ in the form

$$\#3 - (\#11.\#3)(\#11) - (\#17.\#3)(\#17)$$

Simplify on line 19 to $[0, -1/2, 1/2]$. Again we need to normalize this vector so we ask for its length on line 20 and obtain on line 21, $\sqrt{2}/2$.

Finally, we divide the vector by its length to obtain a normalized $\mathbf{u}_3$ on line 23.

Remark

We have used this DERIVE activity to solve a problem. We do not claim that this is the neatest solution, or the shortest. In fact, we have inputted lines we do not use, but we hope that you will, by practice, find a satisfactory solution to any of the problems set.

8.4   SYMMETRIC MATRICES

We introduced the idea of a symmetric matrix in Section 1.5, where we defined a square matrix $A = (a_{ij})$ to be symmetric if A were equal to its transpose, A^T. We now state some facts without proof about symmetric matrices. The proofs can be found in the classical texts on linear algebra. We will show verification by examples.

Facts

Throughout this section we assume that A is an $n \times n$ symmetric matrix.

(i)   All the eigenvalues of A are real.
(ii) A has n linearly independent eigenvectors and thus from our previous work a symmetric matrix can always be diagonalized.
(iii)   Eigenvectors associated with different eigenvalues are always orthogonal to each other.
(iv) A has n orthonormal eigenvectors and hence we can write A as the product PDP^T, where P is orthogonal and D is the diagonal matrix with entries being the eigenvalues. We may need to use the Gram–Schmidt process to ensure that the vectors are orthonormal.

Remember that if P is orthogonal, then $P^{-1} = P^T$ (the transpose), hence we have the equations

$$A = PDP^T = PDP^{-1}$$

DERIVE ACTIVITY 8C

(A)   You will need to load the utility file VECTOR.MTH for this activity.
            Consider the symmetric matrix given below:

$$\begin{bmatrix} 2 & -2 \\ -2 & 5 \end{bmatrix}$$

Use DERIVE to find its eigenvalues and note that they are real and in this case distinct.   (Fact (i).) Then find the corresponding eigenvectors using the EXACT_EIGENVECTOR command as in previous activities.   (Fact (ii).) This case will produce two eigenspaces, each of dimension 1.
            The eigenvectors you write down from the DERIVE screen will probably not be normalized, that is, of unit length, so normalize them. Then, as before, write down the matrix P, which will have the normalized eigenvectors as its columns. Note that these vectors are orthogonal to each other. Take the dot product to check the fact. (Fact (iii).)
            Finally, check that $PDP^T = A$, where D will have the corresponding eigenvalues on its diagonal and zero elsewhere. (Fact (iv).)

Solution

The eigenvalues are 1 and 6.   The corresponding eigenvectors, without fractions, are of the form

$$[2\ 1],\ [1\ -2]$$

Normalized these become

$$\begin{bmatrix} 2/\sqrt{5} & 1/\sqrt{5} \end{bmatrix},\ \begin{bmatrix} 1/\sqrt{5} & -2/\sqrt{5} \end{bmatrix}$$

and their dot product is zero. The matrices P and D are

$$P = \begin{bmatrix} 2/\sqrt{5} & 1/\sqrt{5} \\ 1/\sqrt{5} & -2/\sqrt{5} \end{bmatrix} \qquad D = \begin{bmatrix} 1 & 0 \\ 0 & 6 \end{bmatrix}$$

Finally, check that

$$PDP^T = A \quad \text{and} \quad P^{-1} = P^T$$

(B)   This time take for your matrix A the 3×3 symmetric matrix given below:

$$\begin{bmatrix} 2 & 1 & 1 \\ 1 & 2 & 1 \\ 1 & 1 & 2 \end{bmatrix}$$

We will follow similar steps to part (A) but we will check the answers as we go along.

First find the eigenvalues. DERIVE will give you two, namely, 1 and 4.

As in a previous activity to find which one has multiplicity 2 we can ask for the characteristic polynomial and factorize it. You should find that 1 has multiplicity two.

Again we observe fact (i), that all the eigenvalues are real.

Now ask DERIVE for the eigenvector corresponding to eigenvalue 4. The most obvious vector is of the form $[a \ a \ a]$.

If we choose $a = 1$ we have the vector $[1 \quad 1 \quad 1]$, which when we normalize it will be

$$\begin{bmatrix} 1/\sqrt{3} & 1/\sqrt{3} & 1/\sqrt{3} \end{bmatrix}$$

Thus the eigenspace corresponding to the eigenvalue 4 is one-dimensional and the above normalized vector is a basis for it.

Now we consider the eigenvector(s) corresponding to the eigenvalue 1. DERIVE gives a vector of the form

$$[a \ b \ (-a - b)]$$

We observe that we have two degrees of freedom and therefore we expect a two-dimensional eigenspace.

We can choose any two vectors which satisfy the above form and which are independent of each other. If we choose them so that they are already orthogonal we would not need to use Gram–Schmidt to orthogonalize them. In practice, choosing them orthogonal is not always easy, especially if there are more than two vectors to be found. In this case, therefore, we will proceed to use Gram–Schmidt rather than finding orthogonal vectors by 'inspection'.

Assume we picked the vectors $[0 \ 1 \ -1]$ and $[1 \ 0 \ -1]$.

First we normalize $[0 \ 1 \ -1]$ and obtain the vector

$$\begin{bmatrix} 0 & 1/\sqrt{2} & -1/\sqrt{2} \end{bmatrix}$$

Now we use the Gram–Schmidt process, recalling the equation for $\mathbf{u}_2$ as

$$\lambda\mathbf{u}_2 = \mathbf{v}_2 - (\mathbf{v}_2 \cdot \mathbf{u}_1)\mathbf{u}_1$$

Here $\mathbf{v}_2 = [1 \ \ 0 \ -1]$ and $\mathbf{u}_1 = \begin{bmatrix} 0 & 1/\sqrt{2} & -1/\sqrt{2} \end{bmatrix}$.

By substitution we have

$$\lambda\mathbf{u}_2 = [1 \ -1/2 \ -1/2]$$

By using the ABS command or otherwise we find that $\lambda = \sqrt{3}/\sqrt{2}$, and hence $\mathbf{u}_2$ is given by

$$\begin{bmatrix} \sqrt{2}/\sqrt{3} & -1/\sqrt{6} & -1/\sqrt{6} \end{bmatrix}$$

Fact (iii) said that eigenvectors associated with different eigenvalues are orthogonal, so observe that the eigenvector associated with the eigenvalue 4 is indeed orthogonal to the two eigenvectors we have just produced for the eigenvalue 1.

Now we can either input the matrix P into DERIVE or gather together the three eigenvectors from their respective lines and build up the matrix P. Whichever way we do it the matrices P and D should be

$$P = \begin{bmatrix} 0 & \sqrt{2}/\sqrt{3} & 1/\sqrt{3} \\ 1/\sqrt{2} & -1/\sqrt{6} & 1/\sqrt{3} \\ -1/\sqrt{2} & -1/\sqrt{6} & 1/\sqrt{3} \end{bmatrix} \qquad D = \begin{bmatrix} 1 & 0 & 0 \\ 0 & 1 & 0 \\ 0 & 0 & 4 \end{bmatrix}$$

Finally, use DERIVE to evaluate the product, PDP^T, which should be the same as A. Also, $P^T = P^{-1}$ can be checked in DERIVE. Fact (iv) is now complete and checked.

8.5   QR DECOMPOSITION

In this section we use some of the ideas from preceding sections to obtain a result which is famous in linear algebra and which goes under the name of the QR decomposition.

The Main Result

If A is an $m \times n$ matrix with $m \geq n$, and with all its columns linearly independent, that is, the n columns are all independent, then A can be decomposed into the product of two matrices which we label Q and R, where Q is an $m \times n$ matrix with orthonormal columns and R is an $n \times n$ upper triangular matrix which has an inverse. We write $A = QR$. In the case where $m = n$, the matrix Q will in fact be an orthogonal matrix.

Because Q is stated to have orthonormal columns you might hazard a guess that we take the original matrix A and use its columns in the Gram–Schmidt process to produce Q. That is precisely what we do. The upper triangular matrix comes out in the process. The main result is proved and then some worked examples illustrate the result.

Proof

Consider Section 8.3 above and recall the expressions we obtained there for vectors $\{\mathbf{u}_1, \mathbf{u}_2, \dots, \mathbf{u}_n\}$ which were produced from a basis of vectors $\{\mathbf{v}_1, \mathbf{v}_2, \dots, \mathbf{v}_n\}$:

$$\mathbf{u}_1 = \frac{\mathbf{v}_1}{|\mathbf{v}_1|}$$

$$\lambda \mathbf{u}_2 = \mathbf{v}_2 - (\mathbf{u}_1 \cdot \mathbf{v}_2)\mathbf{u}_1$$

$$\mu \mathbf{u}_3 = \mathbf{v}_3 - (\mathbf{u}_1 \cdot \mathbf{v}_3)\mathbf{u}_1 - (\mathbf{u}_2 \cdot \mathbf{v}_3)\mathbf{u}_2$$

$$\nu \mathbf{u}_4 = \mathbf{v}_4 - (\mathbf{u}_1 \cdot \mathbf{v}_4)\mathbf{u}_1 - (\mathbf{u}_2 \cdot \mathbf{v}_4)\mathbf{u}_2 - (\mathbf{u}_3 \cdot \mathbf{v}_4)\mathbf{u}_3$$

The vectors we are given are the columns of the matrix A, so instead of using $\mathbf{v}_i$ we will use $\mathbf{c}_i$ to remind us that we are handling columns. We use the Gram–Schmidt process to change the columns, $\mathbf{c}_i$, to orthonormal vectors, $\mathbf{u}_i$.

From the first equation above we obtain

$$\mathbf{u}_1 = \frac{\mathbf{c}_1}{|\mathbf{c}_1|}$$

or, rearranged:

$$\mathbf{c}_1 = |\mathbf{c}_1|\mathbf{u}_1$$

From the second equation we have

$$\lambda u_2 = c_2 - (u_1 \cdot c_2)u_1$$

which when rearranged can be written

$$c_2 = \lambda u_2 + (u_1 \cdot c_2)u_1$$

We continue and use the third equation to obtain

$$c_3 = \mu u_3 + (u_1 \cdot c_3)u_1 + (u_2 \cdot c_3)u_2$$

Finally, for our fourth equation we have

$$c_4 = \nu u_4 + (u_1 \cdot c_4)u_1 + (u_2 \cdot c_4)u_2 + (u_3 \cdot c_4)u_3$$

Bringing these together as a system of equations we have

$$c_1 = |c_1| u_1$$
$$c_2 = \lambda u_2 + (u_1 \cdot c_2)u_1$$
$$c_3 = \mu u_3 + (u_1 \cdot c_3)u_1 + (u_2 \cdot c_3)u_2$$
$$c_4 = \nu u_4 + (u_1 \cdot c_4)u_1 + (u_2 \cdot c_4)u_2 + (u_3 \cdot c_4)u_3$$

$$c_n = \omega u_n + (u_1 \cdot c_n)u_1 + (u_2 \cdot c_n)u_2 + (u_3 \cdot c_n)u_3 \ldots + (u_{n-1} \cdot c_n)u_{n-1}$$

This system can now be written as a product:

$$[c_1 \ c_2 \ c_3 \ldots c_n] = [u_1 \ u_2 \ldots u_n] \begin{bmatrix} |c_1| & (u_1 \cdot c_2) & (u_1 \cdot c_3) & (u_1 \cdot c_4) & \ldots & (u_1 \cdot c_n) \\ 0 & \lambda & (u_2 \cdot c_3) & (u_2 \cdot c_4) & \ldots & (u_2 \cdot c_n) \\ 0 & 0 & \mu & (u_3 \cdot c_4) & \ldots & (u_3 \cdot c_n) \\ 0 & 0 & 0 & \nu & \ldots & (u_4 \cdot c_n) \\ \ldots & \ldots & \ldots & \ldots & \ldots & \ldots \\ 0 & 0 & 0 & 0 & \ldots & \omega \end{bmatrix}$$

The first matrix in the product is Q and has the orthonormal vectors which we generated by the Gram–Schmidt process as columns.

The second matrix is R and is certainly upper triangular. The fact that it has an inverse is immediate when you consider the determinant of R, which is the product of the elements on the diagonal, as we saw earlier in the book. The fact that none of the elements on the diagonal can be zero means that the determinant is non-zero and thus an inverse for R exists. We have thus proved our main result for this section.

We will now see how this is put into practice with two examples.

Example 8C

Find the QR decomposition of the 3×2 matrix, A, given below:

$$A = \begin{bmatrix} 1 & 0 \\ 1 & 2 \\ 1 & 1 \end{bmatrix}$$

Remember we can handle $m \times n$ matrices, provided that $m \geq n$.

Solution

Examine the columns and check that they are linearly independent. One is not a multiple of the other so we are sure that A fits the main result's criteria.

Now let $c_1 = [1 \ 1 \ 1]^T$ and $c_2 = [0 \ 2 \ 1]^T$. Then

$$u_1 = c_1 / |c_1| = \left[1/\sqrt{3} \quad 1/\sqrt{3} \quad 1/\sqrt{3} \right]$$

and

$$\lambda u_2 = [0 \quad 2 \quad 1] - \left(\sqrt{3} \right) \left[1/\sqrt{3} \quad 1/\sqrt{3} \quad 1/\sqrt{3} \right]$$

$$= [-1 \quad 1 \quad 0]$$

from which we have

$$u_2 = \left[-1/\sqrt{2} \quad 1/\sqrt{2} \quad 0 \right]$$

Hence we can write down the matrix Q as

$$\begin{bmatrix} 1/\sqrt{3} & -1/\sqrt{2} \\ 1/\sqrt{3} & 1/\sqrt{2} \\ 1/\sqrt{3} & 0 \end{bmatrix}$$

Furthermore, we can write down R from the equations as

$$\begin{bmatrix} \sqrt{3} & \sqrt{3} \\ 0 & \sqrt{2} \end{bmatrix}$$

Finally, using DERIVE or otherwise compute QR and show that $QR = A$.

Example 8D

Find the QR decomposition of the following matrix:

$$\begin{bmatrix} 4 & 1 \\ 3 & -1 \end{bmatrix}$$

Solution

Again we can check that the columns are linearly independent.

Here $c_1 = [4 \ 3]^T$ and $c_2 = [1 \ -1]^T$. Hence,

$$u_1 = (1/5)[4 \ 3] = [4/5 \ 3/5]$$

and

$$\lambda u_2 = [1 \ -1] - (1/5)[4/5 \ 3/5]$$

$$= [21/25 \ -28/25]$$

Using DERIVE or by hand the magnitude of the right-hand side which is λ is equal to 7/5. Thus we have

$$u_2 = [3/5 \ -4/5]$$

The matrices Q and R are then

$$Q = \begin{bmatrix} 4/5 & 3/5 \\ 3/5 & -4/5 \end{bmatrix} \qquad R = \begin{bmatrix} 5 & 1/5 \\ 0 & 7/5 \end{bmatrix}$$

Finally, check that $A = QR$ and that R has an inverse.

Consolidation Exercise

1. Show that if the vector **u** is orthogonal to the vector **v**, then every scalar multiple of **u** is orthogonal to **v**.

    Find a unit vector which is orthogonal to both of the vectors given below, in the vector space $\Re^3$:

    $$v_1 = [1 \ \ 1 \ \ 2] \quad \text{and} \quad v_2 = [0 \ \ 1 \ \ 3]$$

2. Find an orthogonal 3×3 matrix whose first row is

    $$1/\sqrt{3} \quad 1/\sqrt{3} \quad 1/\sqrt{3}$$

    Hint. Remember that every row will be orthogonal to every other row, and that the row vectors all have length one. Check that the row vector given above is a unit vector. Check that in the matrix you produce the columns are also orthogonal to each other and of unit length. Finally, verify, using your matrix, the fact that the determinant of an orthogonal matrix is either +1 or –1.

3. Use the Gram–Schmidt orthogonalization process to change the basis

    $$\{[1 \ \ 1], [2 \ \ 3]\}$$

    in $\Re^2$ to an orthonormal basis in $\Re^2$.

4. Use the Gram–Schmidt orthogonalization process to change the basis

    $$\{[1 \ \ 1 \ \ 1], [1 \ \ 2 \ \ 3], [2 \ \ -1 \ \ 1]\}$$

    in $\Re^3$ to an orthonormal basis in $\Re^3$.

9

Some Applications

9.1   NETWORKS

Figure 9.1 shows a network of pipes.

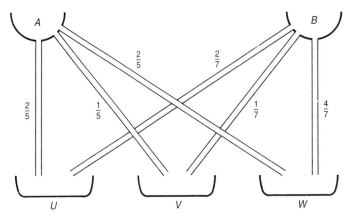

Figure 9.1   A network of pipes connect A and B to U, V and W

In the diagram, we let a and b denote the total amounts of flow entering and leaving A and B respectively. We also let u, v and w denote the total amounts entering at U, V and W respectively.

What are the amounts u, v and w in terms of a and b if the water divides in the fractions shown?

After a little thought you should obtain the following equations:

$$u = 2a/5 + 2b/7$$
$$v = a/5 + b/7$$
$$w = 2a/5 + 4b/7$$

These equations can be written in matrix form as usual. We have

$$\begin{bmatrix} u \\ v \\ w \end{bmatrix} = \begin{bmatrix} 2/5 & 2/7 \\ 1/5 & 1/7 \\ 2/5 & 4/7 \end{bmatrix} \begin{bmatrix} a \\ b \end{bmatrix}$$

From this we identify the matrix which represents this particular network. It is the 3×2 matrix of coefficients.

In this case it is 3×2, because we have three receivers, U, V and W, and two sources, A and B. Turning the network on its side we arrive at Figure 9.2. This is the more usual diagrammatic form of a network. We will assume in this section that the direction of the flows in all our diagrams is from left to right.

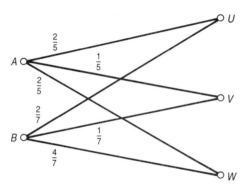

Figure 9.2   A simple network diagram

Consider the matrix of coefficients carefully and try to recall where we have seen this type of matrix before in this book. As a clue, observe that the columns add up to 1 and each entry a_{ij} satisfies the inequality $0 \le a_{ij} \le 1$. Look back to Chapter 4, Section 4.4. This matrix is in fact an example of a column stochastic matrix. When you think about the network from which it comes, if we assume no loss of fluid, then the matrix of coefficients, or the matrix representing the network, will always be column stochastic.

Notation

In networks we call points such as A, B, U, V and W nodes. We call the lines joining nodes either edges or arcs.

Now let us suppose that we enlarge our network by introducing a second network with nodes P and Q. Now u, v and w will be the inputs to the nodes P and Q. See Figure 9.3.

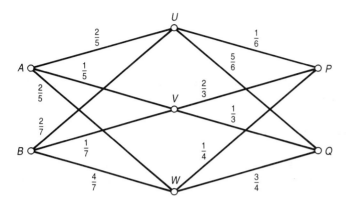

Figure 9.3   A larger network

What are the total inputs at the nodes P and Q in terms of the outputs a and b from the nodes A and B? We already have the equations for the first network as

$$\begin{bmatrix} u \\ v \\ w \end{bmatrix} = \begin{bmatrix} 2/5 & 2/7 \\ 1/5 & 1/7 \\ 2/5 & 4/7 \end{bmatrix} \begin{bmatrix} a \\ b \end{bmatrix}$$

From the new network we have the equations for p and q as

$$\begin{bmatrix} p \\ q \end{bmatrix} = \begin{bmatrix} 1/6 & 2/3 & 1/4 \\ 5/6 & 1/3 & 3/4 \end{bmatrix} \begin{bmatrix} u \\ v \\ w \end{bmatrix}$$

Combining the two equations we arrive at the matrix equation which connects a, b with p, q, namely,

$$\begin{bmatrix} p \\ q \end{bmatrix} = \begin{bmatrix} 1/6 & 2/3 & 1/4 \\ 5/6 & 1/3 & 3/4 \end{bmatrix} \begin{bmatrix} 2/5 & 2/7 \\ 1/5 & 1/7 \\ 2/5 & 4/7 \end{bmatrix} \begin{bmatrix} a \\ b \end{bmatrix}$$

Hence,

$$\begin{bmatrix} p \\ q \end{bmatrix} = \begin{bmatrix} 3/10 & 2/7 \\ 7/10 & 5/7 \end{bmatrix} \begin{bmatrix} a \\ b \end{bmatrix}$$

This equation gives us the result we asked for above, namely, the inputs at P and Q in terms of the outputs at A and B.

Comments on this Application

(i)   Observe that in all the matrices generated above, the columns sum to 1, as would be expected from the original discussion, and thus we have three different column stochastic matrices.

(ii)   If we let

$$A = \begin{bmatrix} 1/6 & 2/3 & 1/4 \\ 5/6 & 1/3 & 3/4 \end{bmatrix}, \quad B = \begin{bmatrix} 2/5 & 2/7 \\ 1/5 & 1/7 \\ 2/5 & 4/7 \end{bmatrix}, \quad C = \begin{bmatrix} 3/10 & 2/7 \\ 7/10 & 5/7 \end{bmatrix}$$

then

$$C = AB$$

Hence, we make the observation that if we combine networks in the way indicated above, we simply need to multiply the relevant matrices to obtain the composite result.

(iii)   Observe that the product matrix, C, is also column stochastic.

Generalisation

The last result is true in general, that is, the product of column stochastic matrices is again stochastic. It follows from considering the transpose of such a matrix that 'column' can be replaced by 'row' in the above and the result will still be true. Hence, a product of doubly stochastic matrices is again doubly stochastic.

Exercise 9A

1.   Write down the matrices corresponding to the networks shown in the following figure.

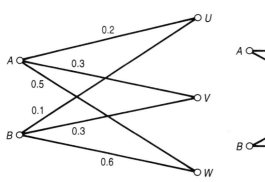
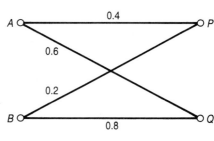

2.   Draw networks which correspond to the following matrices:

(i)
$$\begin{bmatrix} 0.1 & 0.5 & 0.3 \\ 0.2 & 0.5 & 0.6 \\ 0.7 & 0.0 & 0.1 \end{bmatrix}$$

(ii)
$$\begin{bmatrix} 0.4 & 0.2 & 0.7 \\ 0.6 & 0.8 & 0.3 \end{bmatrix}$$

3.   Show that a row (column) stochastic matrix always has an eigenvalue equal to 1. (Again experiment with DERIVE in the 2 × 2 case and then try to generalize.)

4.   If $A = \begin{bmatrix} a & . \\ . & . \end{bmatrix}$ is a 2 × 2 doubly stochastic matrix, fill in the missing three elements.

    Similarly fill in the missing elements for the doubly stochastic 2 × 2 matrix

$$B = \begin{bmatrix} b & . \\ . & . \end{bmatrix}$$

Show that $A \cdot B$ is also doubly stochastic. (You may wish to experiment with DERIVE to tackle this question.) Use DERIVE to consider the 3 × 3 case.

5.   Consider the following network. It can be thought of as two networks put together. One from A, B, C to U, V and the other from U, V to P, Q, R. Write down the matrices for the two separate networks.

    Using matrix multiplication, find the matrix which represents the flow from A, B, C to P, Q, R. Hence, if 10 litres of liquid leaves A, 20 litres leaves B and 30 litres leaves C, use your result to work out the amounts, in litres, that arrive at P, Q and R respectively.

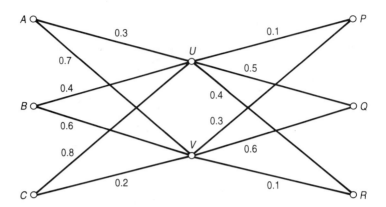

Further applications which are essentially network problems can be found in road systems, leading to graph theory and queuing theory. In physics, applications involving Kirchoff's laws and circuits can be found. In operations research, applications to linear programming and the simplex method can be found.

9.2   STEADY STATE AND MARKOV PROCESSES

In Chapter 4 we introduced the stochastic matrix and gave its alternative name, that is, a probability matrix. In the same way we have the notion of a *probability vector*, which is either a row or column vector with the property that all its entries add up to one.

Definition

If A is a column stochastic $n \times n$ matrix a steady state vector or equilibrium vector for A is an $n \times 1$ column probability vector $\mathbf{v}$, such that $A\mathbf{v} = \mathbf{v}$. If we interpret this definition with the ideas that we have seen so far, this means that the matrix A has an eigenvalue of one.

   If you completed Exercise 9A you will have shown that every probability matrix has an eigenvalue of one and therefore every probability or column stochastic matrix must have a steady state vector.

Application

A traditional application is a migration matrix or a movement of population matrix.

Consider the following idea. The two cities Liverpool and London have a population which moves from one city to the other to live and perhaps work. The following matrix might describe the situation:

To Liverpool London

$$\text{From Liverpool} \begin{bmatrix} 97\% & 1\% \\ 3\% & 99\% \end{bmatrix}$$
$$\text{London}$$

If we write the matrix with decimal entries we will in fact have the column stochastic matrix

$$\begin{bmatrix} 0.97 & 0.01 \\ 0.03 & 0.99 \end{bmatrix}$$

What the matrix means is that 97% of Liverpool's population stays in Liverpool while 1% 'migrates' to London. Also, 3% of London's population comes to live in Liverpool while 99% of London's population stays put!

What is its steady state vector and what does it mean? Using DERIVE or by hand solve the equation $Av = v$.

You should find that $v = [1 \quad 3]^T$. We write this as the equivalent probability vector $[0.25 \ 0.75]^T$.

What this means is that if we begin with the population of London being three times that of Liverpool, then in the long run this ratio will remain approximately the same. Any movement from London to Liverpool will be balanced by movement from Liverpool to London, year-by-year.

Check also that the original matrix has an eigenvalue of 1.

Definition

A probability matrix, P, is said to be *regular* if all the entries of at least one of its powers, P^n, are greater than zero. This means that apart from the columns adding to one, the entries which are already greater than or equal to zero must now be restricted to non-zero values. The migration matrix given above is regular since P does not have any zero entries.

Facts

It is a fact that every regular probability matrix, P, has a unique probability vector, v, such that $Pv = v$. This means that every regular probability matrix has a unique steady state vector. Further, it is a fact that for any probability vector, w, the probability vector, $P^n w$, gets closer and closer to the steady state vector the larger n becomes, where P is a regular probability matrix. This fact gives rise to the expression used above, 'in the long run'.

Finally, as n becomes larger it is a fact that each column of P^n gets closer to the steady state vector.

DERIVE ACTIVITY 9A

(A)   Using the migration matrix given above and DERIVE, show that the columns of P^n tend to the steady state vector. Without DERIVE you would probably never satisfy yourself that this result is true!

Input the matrix P, that is,

$$\begin{bmatrix} 0.97 & 0.01 \\ 0.03 & 0.99 \end{bmatrix}$$

Compute the first few powers of this matrix.

You should find that that the columns are not very close to the steady state vector.

Now, using the power of DERIVE, **Author** P^{100}. Approximate it if you are in exact mode. To two decimal places you should have the matrix

$$\begin{bmatrix} 0.26 & 0.24 \\ 0.73 & 0.75 \end{bmatrix}$$

We are at last getting close!

Now **Author** P^{200} and **Simplify**. DERIVE might take a little time to do this depending on the power of the particular machine you use. The slowest we found to be about five minutes.

The result to two decimal places does indeed give the columns as the steady state vector, that is, P^{200} is the matrix

$$\begin{bmatrix} 0.25 & 0.25 \\ 0.75 & 0.75 \end{bmatrix}$$

Now use DERIVE to show that the second fact is true.

Choose any probability vector, w.

Work out $Pw, P^2 w, \ldots$.

In the long run $P^n w$ should be our steady state vector.

We chose $w = [0.4 \ 0.6]^T$ and our results to two decimal places were

$$P^{10}w = [0.35 \ \ 0.65]^T \quad \text{and} \quad P^{100}w = [0.25 \ \ 0.75]^T$$

For $P^{200}w$, we obtained the matrix $[0.250 \ \ 0.750]^T$ to three decimal places.

Clearly, different probability matrices will tend to their unique steady state vector at different rates.

(B)   As a further example for you to work on, use the matrix

$$P = \begin{bmatrix} 1/2 & 1/3 \\ 1/2 & 2/3 \end{bmatrix}$$

First find its steady state vector by using DERIVE with the command, EXACT_EIGENVECTOR, as we did in Chapter 7. Remember, you need to load the utility file VECTOR.MTH to perform this operation. Furthermore, you need to make the eigenvector, that is, the steady state vector, into a probability vector, before proceeding. Now compute some powers of this new matrix and approximate them. You should find in this case that P^2, P^4 and P^8 suffice to show the steady state vector appearing in the columns, to two or three decimal places. Experiment further with higher powers and more decimal places.

Again, choose any probability vector w and compute P^nw for various values of n as before. You should have the steady state vector to two decimal places quite quickly.

We finish this activity by showing that in general any probability vector, $w = [x \ \ y]^T$, can indeed be chosen, and that for certain powers, n, of the regular probability matrix, P,

$$P^nw = v(\text{the steady state vector})$$

We start by letting $v = [a \ \ b]$, observing that $x + y = 1$.

Further, we observe that for a high enough value of n, P^n will have for its columns the steady state vector, v. Thus we have for some n

$$P^nw = \begin{bmatrix} a & a \\ b & b \end{bmatrix}\begin{bmatrix} x \\ y \end{bmatrix} = \begin{bmatrix} ax+ay \\ bx+by \end{bmatrix} = \begin{bmatrix} a(x+y) \\ b(x+y) \end{bmatrix}$$

$$= \begin{bmatrix} a \\ b \end{bmatrix}$$

since $x + y = 1$. This verifies the result stated.

Further applications involving Markov processes can be found in weather prediction, politics, psychology, game theory, difference equations and differential equations.

9.3   AN APPLICATION FROM ECONOMICS

Suppose that the projected demand for steel in 1980 by the consumer goods industries is 200 million tons. Suppose that the projected demand for coal in the production of steel is 160 million tons. Is this really the amount of coal that needs to be produced? The answer is a resounding NO since other industries besides the steel industry require coal. Similarly, the coal industry will require steel for machines and lorries among other things.

Below we produce a simple model for the economist to ascertain how much steel and coal really needs to be produced to supply all demands.

Demand for steel = final demand for steel + coal industry's demand for steel.

Demand for coal = final demand for coal + steel industry's demand for coal.

Suppose 0.1 tons of steel are needed to produce 1 ton of coal and 0.4 tons of coal are required to produce 1 ton of steel. Let the output of steel be X and the output of coal be Y, in millions of tons. The economist wants to know the values of X and Y. For the steel industry to be in equilibrium we have the equation, from the given data:

$$X = 200 + 0.1Y$$

For the coal industry to be in equilibrium we have a similar equation, namely,

$$Y = 160 + 0.4X$$

Thus we have a system of equations

$$\begin{aligned} X - 0.1Y &= 200 \\ -0.4X + \quad Y &= 160 \end{aligned}$$

In matrix form we have

$$\begin{bmatrix} 1 & -0.1 \\ -0.4 & 1 \end{bmatrix} \begin{bmatrix} X \\ Y \end{bmatrix} = \begin{bmatrix} 200 \\ 160 \end{bmatrix}$$

We have seen how to solve this system in many ways in this book. Here we will use the inverse matrix approach. Thus,

$$\begin{bmatrix} X \\ Y \end{bmatrix} = \frac{1}{0.96}\begin{bmatrix} 1 & 0.1 \\ 0.4 & 1 \end{bmatrix}\begin{bmatrix} 200 \\ 160 \end{bmatrix}$$

$$= \begin{bmatrix} 225 \\ 250 \end{bmatrix}$$

Hence interpreting the result we see that 225 million tons of steel need to be produced by the steel industry, with 200 million tons being needed by the steel industry itself and 25 million tons being needed by the coal industry. Similarly, 250 million tons of coal need to be produced by the coal industry, with 160 million tons being needed by the coal industry itself and 90 million tons being needed by the steel industry.

Tabulated Data

		Steel	Coal	Supply to Final Demand	Total Output
Demand for	Steel	–	25	200	225
	Coal	90	–	160	250

Clearly, this problem is simple in everyday terms but the same idea can be enlarged to many variables and solved using DERIVE or another computer algebra package in a similar way to that given above.

Here is a problem for you to try with three variables instead of two.

Exercise 9B

You are asked, as an economist, to plan the output of coal (X), steel (Y), and lorries, (Z) given the following information.

To produce 1 ton of coal requires 0.4 tons of steel and 0.2 lorries.
To produce 1 ton of steel requires 0.2 tons of coal and 0.1 lorries.
To produce 1 lorry requires 0.6 tons of coal and 0.4 tons of steel.

Consumers demand 10 million tons of coal, 20 million tons of steel and 10 million lorries.

Calculate what should be the total output of coal, steel and lorries.

Tabulate your results to show the supply and demand for the three commodities.

9.4   CODING AND DECODING MESSAGES

Probably one of the simplest codes is the one that takes letters and replaces them with numbers using the rule $A = 1, B = 2, C = 3$, and so on. The message,

  I like linear algebra

now looks like

  9 12 9 11 5 12 9 14 5 1 18 1 12 7 5 2 18 1

We are going to send this message via matrices to a receiver. In this case we will use 2×2 matrices, but any size of square matrix can be used. We put the numbers in the rows of the 2×2 matrices, filling them as we go. Hence we now have the following matrices:

$$\begin{bmatrix} 9 & 12 \\ 9 & 11 \end{bmatrix} \quad \begin{bmatrix} 5 & 12 \\ 9 & 14 \end{bmatrix} \quad \begin{bmatrix} 5 & 1 \\ 18 & 1 \end{bmatrix} \quad \begin{bmatrix} 12 & 7 \\ 5 & 2 \end{bmatrix} \quad \begin{bmatrix} 18 & 1 \\ 0 & 0 \end{bmatrix}$$

Notice that the last matrix only held the two letters $R = 18$ and $A = 1$, and when this occurs we fill the rest of the matrix with zero entries. Now, if we sent these matrices to our receiver exactly as they are, then anyone could 'crack the code' after a short time. What we actually do is send five matrices which are related to the above five.

  We pick a *coding matrix* which, in this case, has to be 2×2 and, most importantly, it must have an inverse. If it didn't, then we would not be able to decode the message accurately. In this example we will choose the coding matrix

$$\begin{bmatrix} 1 & 2 \\ 2 & 5 \end{bmatrix}$$

Its determinant is 1 and thus it has an inverse, which we remind you is unique. We now take each of the five matrices and multiply each of them, on the left, by our coding matrix. If you perform this operation yourself you should obtain the following five matrices respectively:

$$\begin{bmatrix} 27 & 34 \\ 63 & 79 \end{bmatrix} \quad \begin{bmatrix} 23 & 40 \\ 55 & 94 \end{bmatrix} \quad \begin{bmatrix} 41 & 3 \\ 100 & 7 \end{bmatrix} \quad \begin{bmatrix} 22 & 11 \\ 49 & 24 \end{bmatrix} \quad \begin{bmatrix} 18 & 1 \\ 36 & 2 \end{bmatrix}$$

These are the coded matrices which we send to the receiver. Observe that the letter I in the first matrix which was represented by 9 is now represented by 27 and 63. Observe that the two zeros which filled up the fifth matrix now appear as 36 and 2. You may like to observe other facts from this one example. The message is now scrambled and is hard to crack.

We have to send the receiver the coding matrix and the fact that we multiplied each of our matrices on the left. The receiver now acts as the decoder.

To unscramble the message the receiver has to work out the decoding matrix by finding the inverse of the coding matrix. In our case the decoding matrix is

$$\begin{bmatrix} 1 & 2 \\ 2 & 5 \end{bmatrix}^{-1} = \begin{bmatrix} 5 & -2 \\ -2 & 1 \end{bmatrix}$$

Finally, the receiver multiplies each of the matrices received on the left by this decoding matrix. If all the calculations are done correctly the receiver should now have the original five matrices and can obtain the message by using the rule $A = 1$, $B = 2$, and so on.

DERIVE ACTIVITY 9B

Send the same message using 3×3 matrices instead of 2×2. Notice in this case, because the message has exactly 18 letters, that just two 3×3 matrices will suffice and no zero entries will be needed to fill the last one up. For the coding matrix you will need to choose a 3×3 matrix which has an inverse.

This simple application can lead us to study more complicated codes and, in particular, error-correcting codes. These are beyond the scope of this book and would lead us into another branch of algebra.

9.5   QUADRATIC FORMS AND CONIC SECTIONS

The notion of a quadratic equation $ax^2 + bx + c = 0$ should be a familiar one. If we have two variables x_1 and x_2, then the *order two terms* are

$$x_1^2, \quad x_1 x_2, \quad x_2^2$$

In a similar way if we had three variables x_1, x_2 and x_3, the order two terms would be

$$x_1^2, \quad x_2^2, \quad x_3^2, \quad x_1 x_2, \quad x_1 x_3, \quad x_2 x_3$$

In general, if we have n variables, x_1, x_2, ..., x_n, then the order two terms would be of the form $x_i x_j$, where $0 \le i, j \le n$.

Note that we assume the x_i commute so that $x_i x_j = x_j x_i$.

Definition

A *quadratic form in two variables* x_1 and x_2 is defined to be an expression of the form

$$ax_1^2 + bx_1 x_2 + cx_2^2$$

In some texts this form is referred to as the 'principal part of a quadratic form'.
An example of a quadratic form in our sense is

$$4x_1^2 + 3x_1 x_2 + x_2^2$$

Surprisingly, this can be written in matrix form as follows:

$$\begin{bmatrix} x_1 & x_2 \end{bmatrix} \begin{bmatrix} 4 & 3/2 \\ 3/2 & 1 \end{bmatrix} \begin{bmatrix} x_1 \\ x_2 \end{bmatrix}$$

Check this claim by multiplying out the matrix product.
Note the relationships between the entries in the 2×2 matrix and the coefficients of the three terms in the quadratic form. In general, therefore, we can write the quadratic form

$$ax_1^2 + bx_1 x_2 + cx_2^2$$

in matrix form as

$$\begin{bmatrix} x_1 & x_2 \end{bmatrix} \begin{bmatrix} a & b/2 \\ b/2 & c \end{bmatrix} \begin{bmatrix} x_1 \\ x_2 \end{bmatrix}$$

Notice that if we start with a quadratic form as defined above, the 2×2 matrix obtained in the matrix form must always be symmetrical. Similarly for three variables, we could write down a symmetric 3×3 matrix, as is shown below. A quadratic form in three variables is

$$ax_1^2 + bx_2^2 + cx_3^2 + dx_1 x_2 + ex_1 x_3 + fx_2 x_3$$

In matrix form it can be written as

$$\begin{bmatrix} x_1 & x_2 & x_3 \end{bmatrix} \begin{bmatrix} a & d/2 & e/2 \\ d/2 & b & f/2 \\ e/2 & f/2 & c \end{bmatrix} \begin{bmatrix} x_1 \\ x_2 \\ x_3 \end{bmatrix}$$

You should be able to generalize this idea to quadratic forms in more variables and also see the patterns that emerge to fill in the entries of the square matrix. In all these examples of quadratic forms we have been able to write a matrix form equivalent. In general, the matrix form is

$$\mathbf{x}^T A \mathbf{x}$$

where $\mathbf{x}^T = [x_1 \ x_2 \ x_3 \ ... \ x_n]$ is the transpose of $\mathbf{x}$, and A is a symmetrical $n \times n$ matrix.

In this section we will stay with the two-variable case and thus we will consider 2×2 matrices for A in what follows. You might be familiar with conic sections from geometry but for completeness we revise the ideas we need.

If we consider an upright cone and a plane intersecting the cone we can obtain various sections. We assume that the plane intersects the cone at least at one point. If the plane is parallel to the base, then the section obtained will be a circle or a point if the plane just touches the apex of the cone. If the plane is vertical to the base, then we can obtain the hyperbola. If the plane is parallel to one of the sloping edges, then we obtain the parabola. If the plane intersects the cone at an angle other than those indicated in the preceding sections, then we obtain the ellipse.

The conic sections therefore consist of the shapes,

circle, ellipse, hyperbola and parabola

We ignore the so-called degenerate sections in this application. We also ignore the parabola in what follows.

The equation of a circle, centre the origin and radius r, is given by the equation

$$x_1^2 + x_2^2 = r^2$$

The equation of the ellipse, centre the origin with semi-major axis 'a' and semi-minor axis 'b', is given by

$$\frac{x_1^2}{a^2} + \frac{x_2^2}{b^2} = 1$$

The equation of the hyperbola, centre the origin, is given by

$$\frac{x_1^2}{a^2} - \frac{x_2^2}{b^2} = 1$$

We are now in a position to show the connection between conic sections and quadratic forms. We do this by example.

Example 9A

Write the quadratic form $x_1^2 - 4x_1x_2 + 3x_2^2$ in the form $x^T Ax$.

Write the matrix A in the form PDP^T where P is an orthogonal matrix and D is a diagonal matrix. Identify the conic section that is given by the equation

$$x_1^2 - 4x_1x_2 + 3x_2^2 = 6$$

Solution

$$x_1^2 - 4x_1x_2 + 3x_2^2 = \begin{bmatrix} x_1 & x_2 \end{bmatrix} \begin{bmatrix} 1 & -2 \\ -2 & 3 \end{bmatrix} \begin{bmatrix} x_1 \\ x_2 \end{bmatrix}$$

Hence A is the 2×2 matrix

$$\begin{bmatrix} 1 & -2 \\ -2 & 3 \end{bmatrix}$$

From Section 8.4 we saw that any symmetric matrix can be written in the form PDP^T, where P is an orthogonal matrix and D is a diagonal matrix. To obtain this form we have to find the eigenvalues and the corresponding eigenvectors of A. Use DERIVE to verify that the eigenvalues are exactly, $2 - \sqrt{5}$ and $2 + \sqrt{5}$.

We now need to find the exact eigenvectors for these eigenvalues and then normalize them. Use DERIVE to verify that the eigenvectors when normalized are, respectively,

$$1/(\sqrt{(10-2\sqrt{5})})[2, \quad \sqrt{5}-1]$$

and

$$1/(\sqrt{(10+2\sqrt{5})})[2, \quad -\sqrt{5}-1]$$

Observe or verify that these two vectors are orthogonal.   Note that the vectors $\begin{bmatrix} 2, & \sqrt{5}-1 \end{bmatrix}$ and $\begin{bmatrix} 2, & -\sqrt{5}-1 \end{bmatrix}$ have zero scalar product.   We can write down the matrix P which has the above vectors as its columns.   We then have for P the following 2×2 matrix:

$$\begin{bmatrix} 2/(\sqrt{(10-2\sqrt{5})}) & 2/(\sqrt{(10+2\sqrt{5})}) \\ (\sqrt{5}-1)/(\sqrt{(10-2\sqrt{5})}) & -(\sqrt{5}+1)/(\sqrt{(10+2\sqrt{5})}) \end{bmatrix}$$

Now compute the product PDP^T and show that it is indeed A despite all the irrational numbers! Hence we have diagonalized A and can rewrite the equation

$$PDP^T = A$$

as

$$P^T AP = D$$

since P being orthogonal implies $PP^T = P^T P = I$.

In Section 8.1 we showed that the determinant of an orthogonal matrix had to be $+1$ or -1. Our matrix for P has determinant -1. Check it using DERIVE.

If we interchange the columns we will obtain a matrix with determinant $+1$ by a property of determinants shown in Chapter 2. We can do this interchange since it was our choice as to which column we placed the normalized eigenvectors in. Furthermore, we showed in Section 8.1 that an orthogonal matrix with a determinant of $+1$ was always a rotation. Hence P with its columns interchanged represents a rotation through an angle θ, where $0 \le \theta < 2\pi$.

The last part of the question asked us to identify the conic section represented by

$$x_1^2 - 4x_1 x_2 + 3x_2^2 = 6$$

We can now change the variables from x_1 and x_2 to x and y by rotating whatever the conic section is through the appropriate angle θ so that the mixed term $-4x_1 x_2$ disappears. We then obtain the equation

$$(2-\sqrt{5})x^2 + (2+\sqrt{5})y^2 = 6$$

Equivalently, we can write this equation as

$$\frac{x^2}{6(2+\sqrt{5})}+\frac{y^2}{6(2-\sqrt{5})}=-1$$

which can then be written as

$$\frac{y^2}{6(\sqrt{5}-2)}-\frac{x^2}{6(\sqrt{5}+2)}=1$$

Referring to the beginning of this application we see this is the equation of a hyperbola. We have left out some theory from the last part of this application which we now give with reference to this example. Another way of giving the quadratic form in this exercise is

$$\left[\begin{bmatrix}1 & -2\\ -2 & 3\end{bmatrix}\begin{bmatrix}x_1\\ x_2\end{bmatrix}\right]\cdot\begin{bmatrix}x_1\\ x_2\end{bmatrix}$$

Check that this expression, when evaluated, does indeed give us our original quadratic form. In general, we have that a quadratic form can be written as the scalar product

$$(A\mathbf{v})\cdot\mathbf{v}$$

where A is an $n \times n$ matrix and $\mathbf{v}$ is an $n \times 1$ column vector.
     In Example 9B above, A is

$$\begin{bmatrix}1 & -2\\ -2 & 3\end{bmatrix}$$

and $\mathbf{v}^T$ is the 1×2 vector

$$[x_1 \ x_2]$$

Now we have shown that A can be written as PDP^T so the scalar product in general is

$$(PDP^T\mathbf{v})\cdot\mathbf{v}$$

It is not difficult to prove that, in general,

$$B\mathbf{v}\cdot\mathbf{w} = \mathbf{v}\cdot B^T\mathbf{w}$$

Applying this to $(PDP^T v) \cdot v$, with $B = P$ and $DP^T v$ replacing v and v replacing w, we have

$$P(DP^T v) \cdot v = DP^T v \cdot P^T v$$

Hence our quadratic form can be written as

$$DP^T v \cdot P^T v$$

Alternatively, we can write this as

$$Dz \cdot z$$

where $z = P^T v$.

We claim that this last expression is another quadratic form, for, in our example, D was the 2×2 matrix

$$\begin{bmatrix} 2-\sqrt{5} & 0 \\ 0 & 2+\sqrt{5} \end{bmatrix}$$

and if $z^T = [x \ \ y]$, we have, for $Dz \cdot z$,

$$\begin{bmatrix} 2-\sqrt{5} & 0 \\ 0 & 2+\sqrt{5} \end{bmatrix} \begin{bmatrix} x \\ y \end{bmatrix} \cdot \begin{bmatrix} x \\ y \end{bmatrix}$$

This in turn works out to

$$(2-\sqrt{5})x^2 + (2+\sqrt{5})y^2$$

which is a quadratic form without the mixed term, and is the expression given earlier when we put it equal to 6 and found the resulting equation to be that of a hyperbola. We have therefore verified our claims. In the literature this result is usually known as

The Principal Axis Theorem

This states that if we are given a quadratic form

$$ax_1^2 + bx_1 x_2 + cx_2^2 = m$$

then there exists a number θ such that $0 \leq \theta < 2\pi$ and such that the above quadratic form can be written in the form

$$dx^2 + ey^2 = m$$

where x and y are the new axes obtained by rotating the x_1 and x_2 axes through an angle θ counterclockwise about the origin, and where d and e are the eigenvalues of the matrix of the original quadratic form. We have actually proved this theorem in this section.

9.6   DIFFERENTIAL EQUATIONS

Note that this application needs some understanding of the Calculus.

A first-order ordinary differential equation (ODE) is an equation which contains derivatives to the first order only. Examples of such equations are

$$\frac{dy}{dx} + \sin(x) = 0 \quad \text{and} \quad x^2 \frac{dy}{dx} - \cos(y) = 5x$$

In a similar way a second-order ordinary differential equation is an equation which contains derivatives of the second, and possibly first, order only. Examples of such equations are

$$\frac{d^2y}{dx^2} + 5\frac{dy}{dx} - 7 = 0 \quad \text{and} \quad \frac{d^2y}{dx^2} - 4x = 0$$

Similarly for higher-order differentials.

For completeness we use the word 'ordinary' to distinguish this type of differential equation from 'partial' differential equations (PDE). In this application we will consider first-order ODEs only.

In biology and economics, rates of growth and decay are considered for different reasons. The underlying mathematics is essentially the same. The rate of change of a variable y with respect to x is dy/dx and hence it is not surprising that first-order differential equations model some situations that arise in subjects such as biology and economics.

Consider the situation where the rate of growth is dx/dt, where x is some function of t (time). One of the simplest differential equations is

$$\frac{dx}{dt} = ax(t)$$

This can be interpreted as 'the rate of change of x with respect to time is proportional to the function $x(t)$'. The general solution to this differential equation is

$$x(t) = Ce^{at}$$

where C is an arbitrary constant.

We can check that this is so by differentiating the above equation to get back to the differential equation, that is,

$$\frac{dx}{dt} = Cae^{at} = ax(t)$$

Now let us consider two such differential equations which we wish to solve simultaneously:

$$\frac{dx_1}{dt} = a_{11}x_1(t) + a_{12}x_2(t)$$

and

$$\frac{dx_2}{dt} = a_{21}x_1(t) + a_{22}x_2(t)$$

Using the notation given above you should see that this system of first-order differential equations can be written in matrix form as follows:

$$\begin{bmatrix} \dfrac{dx_1}{dt} \\[2ex] \dfrac{dx_2}{dt} \end{bmatrix} = \begin{bmatrix} a_{11} & a_{12} \\ a_{21} & a_{12} \end{bmatrix} \begin{bmatrix} x_1 \\ x_2 \end{bmatrix}$$

If we now write $\mathbf{x}^T = [x_1 \ x_2]$ as usual, we can rewrite the above equation as

$$\frac{d\mathbf{x}}{dt} = A\mathbf{x}$$

where A is our general 2×2 matrix.

Clearly, we could write down an $n \times n$ system in general and the matrix equation

$$\frac{dx}{dt} = Ax$$

would represent it, with A being an $n \times n$ matrix and x being an $n \times 1$ matrix.

So what is the solution to this matrix differential equation?

If we take our lead from the simple differential equation we started with, the solution there was of the form Ce^{at}, so is there an equivalent solution for our matrix differential equation? It would have to be something like

$$Ce^{At}$$

where A is a square matrix of some size. This leads us to consider the exponential function e^A, with A being a square matrix.

In analysis the exponential function is given in many ways, all of which are equivalent. One such way is in the form of an infinite series

$$e^x = 1 + x + x^2/2! + x^3/3! + \dots$$

where $n!$ is n factorial, that is, $n!$ is the product

$$n(n-1)(n-2) \dots (3)(2)(1)$$

Definition

We define e^A as an $n \times n$ matrix which is given by the formula

$$e^A = I + A + A^2/2! + A^3/3! + \dots$$

where A is an $n \times n$ matrix and I is the $n \times n$ identity matrix.

With this definition we can proceed to solve the matrix differential equation that we introduced above. We consider the possible solution

$$x(t) = e^{At}C$$

where x is our $n \times 1$ vector and C must now be an $n \times 1$ vector for conformability. Note that we cannot write Ce^{At} since e^{At} is $n \times n$ and C would also have to be $n \times n$ but then Ce^{At} would then be an $n \times n$ matrix and not an $n \times 1$ matrix, which $x(t)$ definitely is. To compute the matrix e^{At} in a particular problem can be difficult. Suffice it to say that we can make life a little simpler by accepting the following facts without proof.

If we can diagonalize A to a matrix D, such that $P^{-1}AP = D$, then we only have to evaluate e^{Dt} which is

$$\text{diag}\left[e^{\lambda_1 t}, \ e^{\lambda_2 t}, \ \ldots, \ e^{\lambda_n t}\right]$$

where the λs are the distinct eigenvalues of A. Then

$$P^{-1}e^{At}P = e^{Dt}$$

or, equivalently,

$$e^{At} = Pe^{Dt}P^{-1}$$

We observe that eigenvalues will again play a leading role in this application.

Exercise 9C

1.   Write the following systems of differential equations in matrix form:

(i) $\dfrac{dx_1}{dt} = 3x_1 - 4x_2$ (ii) $\dfrac{dx_1}{dt} = x_2$

   $\dfrac{dx_2}{dt} = -x_1 - 2x_2$ $\dfrac{dx_2}{dt} = 4x_1$

2.   For the following matrices write down the first four terms of e^A and simplify

(i) $A = \begin{bmatrix} 1 & -1 \\ 0 & 2 \end{bmatrix}$ (ii) $A = \begin{bmatrix} 5 & 6 \\ 2 & 1 \end{bmatrix}$

We now give a DERIVE activity which will show how this theory goes in practice. We will solve a 2×2 system of first-order differential equations, observing that bigger systems follow the same method of solution.

DERIVE ACTIVITY 9C

(A)   Consider the simultaneous first-order ordinary differential equations given
       below:

$$\frac{dx_1}{dt} = 5x_1 + 6x_2$$

$$\frac{dx_2}{dt} = 2x_1 + x_2$$

where x_1 and x_2 are assumed to be functions of time, t. Such a system could occur in a
predator–prey model. The equivalent matrix system is given by

$$\begin{bmatrix} \dfrac{dx_1}{dt} \\ \\ \dfrac{dx_2}{dt} \end{bmatrix} = \begin{bmatrix} 5 & 6 \\ 2 & 1 \end{bmatrix} \begin{bmatrix} x_1 \\ x_2 \end{bmatrix}$$

Using DERIVE show that the eigenvalues are real and distinct and have values -1
and 7. Next find the exact eigenvectors for these eigenvalues and show that they can
be represented by $[1\ -1]$ and $[3\ 1]$ respectively. Now let P be the matrix obtained by
putting the eigenvectors as columns. Hence, we have

$$P = \begin{bmatrix} 1 & 3 \\ -1 & 1 \end{bmatrix} \quad \text{and} \quad D = \begin{bmatrix} -1 & 0 \\ 0 & 7 \end{bmatrix}$$

Check that $P^{-1}AP = D$. Hence, $A = PDP^{-1}$ and

$$e^{At} = Pe^{Dt}P^{-1}$$

Now compute e^{At}.
       DERIVE should give you the following matrix:

$$\frac{1}{4}\begin{bmatrix} e^{-t} + 3e^{7t} & -3e^{-t} + 3e^{7t} \\ -e^{-t} + e^{7t} & 3e^{-t} + e^{7t} \end{bmatrix}$$

Finally since $\mathbf{x} = e^{At}\mathbf{C}$, where $\mathbf{C}^T = [C_1 \ C_2]$, we can write down the solutions for x_1 and x_2 as

$$x_1 = \frac{e^{-t}}{4}(C_1 - 3C_2) + \frac{e^{7t}}{4}(3C_1 + 3C_2)$$

$$x_2 = \frac{e^{-t}}{4}(-C_1 + 3C_2) + \frac{e^{7t}}{4}(C_1 + C_2)$$

If we let $C_1 - 3C_2 = 4a$ and $C_1 + C_2 = 4b$ these expressions reduce to

$$x_1 = ae^{-t} + 3be^{7t}$$

and

$$x_2 = -ae^{-t} + be^{7t}$$

Check that these expressions do satisfy the pair of differential equations we started with in this activity by direct differentiation, using DERIVE or by hand.

Finally, we finish this activity with an observation.

Consider again the two solutions that we have just obtained:

$$x_1 = ae^{-t} + 3be^{7t}$$

and

$$x_2 = -ae^{-t} + be^{7t}$$

These can be written in another matrix form, namely,

$$\begin{bmatrix} x_1 \\ x_2 \end{bmatrix} = ae^{-t}\begin{bmatrix} 1 \\ -1 \end{bmatrix} + be^{7t}\begin{bmatrix} 3 \\ 1 \end{bmatrix}$$

Check that this matrix equation does give us the two solutions correctly and then observe that the column matrices on the right-hand side are simply the eigenvectors corresponding to the eigenvalues that are in the indices of the exponential function multiplying them.

As you may imagine, this is not pure coincidence. However, we can only use this simpler method if the eigenvalues are real and distinct, as was the case in this DERIVE activity. The theory for this latter case is quite straightforward. Instead of introducing the exponential matrix, we could have suggested that a solution to the system

$$\begin{bmatrix} \dfrac{dx_1}{dt} \\[2ex] \dfrac{dx_2}{dt} \end{bmatrix} = \begin{bmatrix} a_{11} & a_{12} \\ a_{21} & a_{22} \end{bmatrix} \begin{bmatrix} x_1 \\ x_2 \end{bmatrix}$$

might be of the form, $x_1 = me^{\lambda t}$ and $x_2 = ne^{\lambda t}$ coming directly from considering the solution to the equation $dx/dt = ax(t)$.

Then

$$\frac{dx_1}{dt} = m\lambda e^{\lambda t} \quad \text{and} \quad \frac{dx_2}{dt} = n\lambda e^{\lambda t}$$

and

$$\begin{bmatrix} \dfrac{dx_1}{dt} \\[2ex] \dfrac{dx_2}{dt} \end{bmatrix} = \begin{bmatrix} m\lambda e^{\lambda t} \\ n\lambda e^{\lambda t} \end{bmatrix} = \lambda e^{\lambda t} \begin{bmatrix} m \\ n \end{bmatrix}$$

and

$$\begin{bmatrix} x_1 \\ x_2 \end{bmatrix} = e^{\lambda t} \begin{bmatrix} m \\ n \end{bmatrix}$$

Thus,

$$\begin{bmatrix} \dfrac{dx_1}{dt} \\[2ex] \dfrac{dx_2}{dt} \end{bmatrix} = \begin{bmatrix} a_{11} & a_{12} \\ a_{21} & a_{22} \end{bmatrix} \begin{bmatrix} x_1 \\ x_2 \end{bmatrix}$$

becomes

$$\lambda e^{\lambda t} \begin{bmatrix} m \\ n \end{bmatrix} = A e^{\lambda t} \begin{bmatrix} m \\ n \end{bmatrix}$$

and finally,

$$\lambda \begin{bmatrix} m \\ n \end{bmatrix} = A \begin{bmatrix} m \\ n \end{bmatrix}$$

This equation should by now be looking familiar! It is the eigenvalue equation $Ax = \lambda x$ with $x^T = [m \ n]$. Thus finding the eigenvalues with their corresponding eigenvectors will indeed allow us to write down the solution to our system of ordinary differential equations.

We end this DERIVE activity with a problem that extends the ideas introduced above.

(B)   Solve the following system of ordinary first-order differential equations:

$$\frac{dx_1}{dt} = x_1 - x_2 + 4x_3$$

$$\frac{dx_2}{dt} = 3x_1 + 2x_2 - x_3$$

$$\frac{dx_3}{dt} = 2x_1 + x_2 - x_3$$

with the initial conditions $x_1(0) = 1$, $x_2(0) = 2$, $x_3(0) = 3$.

Solution

In this case the matrix A, which gives us all the information we need, is the matrix of coefficients

$$\begin{bmatrix} 1 & -1 & 4 \\ 3 & 2 & -1 \\ 2 & 1 & -1 \end{bmatrix}$$

Use DERIVE to find the eigenvalues and verify that they are 1, –2, and 3. Since they are real and distinct our 'short-cut' will give us the general solution.

Now find the respective exact eigenvectors and verify that they are

$$[1\ -4\ -1],\quad [1\ -1\ -1],\quad [1\ 2\ 1]$$

Hence, we can write down the general solution as

$$\begin{bmatrix} x_1 \\ x_2 \\ x_3 \end{bmatrix} = C_1 e^{t} \begin{bmatrix} 1 \\ -4 \\ -1 \end{bmatrix} + C_2 e^{-2t} \begin{bmatrix} 1 \\ -1 \\ -1 \end{bmatrix} + C_3 e^{3t} \begin{bmatrix} 1 \\ 2 \\ 1 \end{bmatrix}$$

In this activity we have added an extra line, namely, the initial conditions that this system is working to. This means that at time $t = 0$ the x_1 has to have the value 1 and similarly for the other two functions x_2 and x_3.

Putting this information into our general solution we will arrive at a *particular solution*.

Inserting the initial values gives us the following:

$$\begin{bmatrix} 1 \\ 2 \\ 3 \end{bmatrix} = C_1 \begin{bmatrix} 1 \\ -4 \\ -1 \end{bmatrix} + C_2 \begin{bmatrix} 1 \\ -1 \\ -1 \end{bmatrix} + C_3 \begin{bmatrix} 1 \\ 2 \\ 1 \end{bmatrix}$$

Now we have to find the values of the constants, C_1, C_2 and C_3 to find our particular solution. This entails solving three equations in three unknowns, which we have done many times before in this book. We will use the inverse matrix method in this case.

The system of equations we have to solve is

$$\begin{bmatrix} 1 \\ 2 \\ 3 \end{bmatrix} = \begin{bmatrix} 1 & 1 & 1 \\ -4 & -1 & 2 \\ -1 & -1 & 1 \end{bmatrix} \begin{bmatrix} C_1 \\ C_2 \\ C_3 \end{bmatrix}$$

Using DERIVE input the matrix of coefficients for the C_is. Show that the determinant of your matrix is 6 and therefore has an inverse. Compute the inverse of the matrix of coefficients and hence find the solutions for the C_is.

Verify that the solutions are

$$C_1 = 1, \quad C_2 = -2, \quad C_3 = 2$$

Hence the particular solution is

$$x_1(t) = e^t - 2e^{-2t} + 2e^{3t}$$
$$x_2(t) = -4e^t + 2e^{-2t} + 4e^{3t}$$
$$x_3(t) = -e^t + 2e^{-2t} + 2e^{3t}$$

Again you can check that this is a particular solution of the system given by direct differentiation using DERIVE or by hand.

Answers to Exercises and Solutions to Consolidation Exercises

Chapter 1

Exercise 1A

1. R is 2×2, S is 3×1, T is 3×4.

2. $d_{11} = 1, d_{12} = 2, d_{13} = 3, d_{14} = 4, d_{21} = 5, d_{22} = 7, d_{23} = 0, d_{24} = 11$.

3. $e_{12} + f_{12} = 6 + -1 = 5$
   $e_{21} + f_{21} = 1 + 5 = 6$
   $e_{22} + f_{22} = -3 + 2 = -1$

4. $$\begin{bmatrix} a_{11} & a_{12} & a_{13} & a_{14} \\ a_{21} & a_{22} & a_{23} & a_{24} \\ a_{31} & a_{32} & a_{33} & a_{34} \end{bmatrix}$$

5. $$\begin{bmatrix} a_{11} & a_{12} & a_{13} \\ a_{21} & a_{22} & a_{23} \\ a_{31} & a_{32} & a_{33} \end{bmatrix}$$

DERIVE Activity 1B

The condition for A and A' to be conformable under addition is that they must be square.

Exercise 1B

1. $-2A = \begin{bmatrix} 2 & 6 \\ 4 & -8 \end{bmatrix}$, $\dfrac{A}{2} = \begin{bmatrix} -\frac{1}{2} & -\frac{3}{2} \\ -1 & 2 \end{bmatrix}$

2. $-A = \begin{bmatrix} 1 & 3 \\ 2 & -4 \end{bmatrix}$, $A + -A = \begin{bmatrix} -1 & -3 \\ -2 & 4 \end{bmatrix} + \begin{bmatrix} 1 & 3 \\ 2 & -4 \end{bmatrix} = \begin{bmatrix} 0 & 0 \\ 0 & 0 \end{bmatrix}$

3. Yes, $B = -A$.

4. $2P - 3R = \begin{bmatrix} 2 & -2 \\ 6 & 4 \end{bmatrix} - \begin{bmatrix} 0 & 18 \\ 12 & -15 \end{bmatrix} = \begin{bmatrix} 2 & -20 \\ -6 & 19 \end{bmatrix}$

$0.7P + 0.2R = \begin{bmatrix} 0.7 & -0.7 \\ 2.1 & 1.4 \end{bmatrix} + \begin{bmatrix} 0 & 1.2 \\ 0.8 & -1 \end{bmatrix} = \begin{bmatrix} 0.7 & 0.5 \\ 2.9 & 0.4 \end{bmatrix}$

Exercise 1C

14. (i), (ii) done.

(iii) AB is of size 3×3.
BA is of size 2×2.

(iv) AB is of size 3×4.
BA is impossible.

(v) Neither AB or BA are possible.

DERIVE Activity 1C

(i) $A + B = \begin{bmatrix} 5 & -1 & -1 \\ -1 & 1 & 7 \end{bmatrix}$

(ii) Not possible

(iii) $3B = \begin{bmatrix} 12 & 0 & -9 \\ -3 & -6 & 9 \end{bmatrix}$

(iv) Not possible

(v) $BC = \begin{bmatrix} 11 & -12 & 0 & -5 \\ -15 & 5 & 2 & 4 \end{bmatrix}$

(vi) Not possible

(vii) Not possible

(viii) $AD = \begin{bmatrix} 9 \\ 9 \end{bmatrix}$

(ix) $D^T D = [14]$

(x) $DD^T = \begin{bmatrix} 4 & -2 & 6 \\ -2 & 1 & -3 \\ 6 & -3 & 9 \end{bmatrix}$ (xi)   Not possible

(xii) $B - A = \begin{bmatrix} 3 & 1 & -5 \\ -1 & -5 & -1 \end{bmatrix}$ (xiii)   Not possible

DERIVE Activity 1D

Let $A = \begin{bmatrix} a & b \\ c & d \end{bmatrix}$. A has an inverse if $ad \neq bc$. A hasn't got an inverse if $ad = bc$.

Consolidation Exercise

1.   (i)   Impossible since A and B are not the same size and therefore not conformable under matrix addition.

(ii) AB is possible since A is 2×1 and B is 1×2, therefore we expect a 2×2 matrix for the product.

$$AB = \begin{bmatrix} 3 & 4 \\ 6 & 8 \end{bmatrix}$$

(iii) BA is possible and we would now expect a 1×1 matrix, i.e. a real number.

$$BA = [11]$$

(iv) $2B - 3A$ is not possible since B and A are not conformable for subtraction for a similar reason to (i) above.

(v) $(AB)'$ is immediate from (ii) above.

$$(AB)' = \begin{bmatrix} 3 & 6 \\ 4 & 8 \end{bmatrix}$$

(vi) $B'A' = (AB)'$ in general and therefore the answer for (v) above is the answer to this question also.

N.B. This answers question 3, the first part.

2. (i) CD is not possible since C is 3×3 and D is 1×3 and therefore not conformable under multiplication for this particular product. Observe however that DC is possible to compute. The answer would be a 1×3 matrix.

(ii) C' is possible and is,

$$\begin{bmatrix} -2 & 3 & 4 \\ -1 & 0 & -3 \\ 5 & 1 & 0 \end{bmatrix}$$

(iii) D' is possible and is

$$\begin{bmatrix} -1 \\ 2 \\ -3 \end{bmatrix}$$

(iv) $(DC)'$ is possible provided DC is possible. As stated in (i) above CD can be calculated and thus its transpose can be written down.

$$(DC)' = \begin{bmatrix} -4 & 10 & -3 \end{bmatrix}' = \begin{bmatrix} -4 \\ 10 \\ -3 \end{bmatrix}$$

(v) $C'D' = (DC)'$ as indicated above in question 1 and $C'D'$ has the same answer as given in (iv) above.

N.B. This answers question 3 the second part.

(vi) $C^2 = \begin{bmatrix} 21 & -13 & -11 \\ -2 & 6 & 15 \\ -17 & -4 & 17 \end{bmatrix}$

3. The answer to this question has been given above in the answers to question 1 and 2.

In general if A and B are conformable under multiplication, then $A'B' = (BA)'$. This extends to 3 and more matrices as long as the appropriate product is conformable, i.e. $(ABC)' = C'B'A'$ and so on.

4. (i) $\lambda E = \begin{bmatrix} -6 & 12 \\ 18 & -15 \end{bmatrix}$

(ii) $E - \lambda I = \begin{bmatrix} -2-\lambda & 4 \\ 6 & -5-\lambda \end{bmatrix} = \begin{bmatrix} -5 & 4 \\ 6 & -8 \end{bmatrix}$

(iii) $5\lambda E^2 = 5\lambda \begin{bmatrix} 28 & -28 \\ -42 & 49 \end{bmatrix} = \begin{bmatrix} 140\lambda & -140\lambda \\ -210\lambda & 245\lambda \end{bmatrix} = \begin{bmatrix} 420 & -420 \\ -630 & 735 \end{bmatrix}$

(iv) $5(\lambda E)^2 = 5 \begin{bmatrix} 28\lambda^2 & -28\lambda^2 \\ -42\lambda^2 & 49\lambda^2 \end{bmatrix} = \begin{bmatrix} 140\lambda^2 & -140\lambda^2 \\ -210\lambda^2 & 245\lambda^2 \end{bmatrix} = \begin{bmatrix} 1260 & -1260 \\ -1890 & 2205 \end{bmatrix}$

(v) $5\lambda^2 E^2 = 5\lambda^2 \begin{bmatrix} 28 & -28 \\ -42 & 49 \end{bmatrix} = \begin{bmatrix} 140\lambda^2 & -140\lambda^2 \\ -210\lambda^2 & 245\lambda^2 \end{bmatrix} = \begin{bmatrix} 1260 & -1260 \\ -1890 & 2205 \end{bmatrix}$

From this exercise we have shown that $5(\lambda E)^2 = 5\lambda^2 E^2$.
In general we have $(\alpha E)^n = \alpha^n E^n$ where α is any scalar.
In (iii) we have that $5\lambda E^2 = 5\lambda(E^2)$ which is not the same as (iv) or (v).

5. If $(I - A)$ has an inverse we have to find a matrix B with the property that $(I - A)B = B(I - A) = I$ from the definition of multiplicative inverse.

Consider the hint in the exercise.
If x and y are real numbers we have the following factorizations,

$$x^2 - y^2 = (x - y)(x + y)$$

and $$x^3 - y^3 = (x - y)(x^2 + xy + y^2)$$

and in general

$$x^n - y^n = (x - y)(x^{n-1} + x^{n-2}y + \ldots + xy^{n-2} + y^{n-1})$$

However when we consider matrices we have to beware!

$AB \neq BA$ in general.

Hence $A^2 - B^2 \neq (A - B)(A + B)$ unless ...

If $B = I$ then $AB = BA$ for every square matrix A and the factorizations we produced above for x and y will now hold.

Thus with $B = I$ we have from the general expression above,

$$A^n - I^n = (A - I)(A^{n-1} + A^{n-2}I + \ldots + AI^{n-2} + I^{n-1})$$

which simplifies to,

$$A^n - I = (A - I)(A^{n-1} + A^{n-2} + \ldots + A + I)$$

Now let $n = m + 1$ so that we can use the fact that $A^{m+1} = 0$.

Thus we have the following true equation,

$$A^{m+1} - I = (A - I)(A^m + A^{m-1} + \ldots + A + I)$$

which reduces to

$$-I = (A - I)(A^m + A^{m-1} + \ldots + A + I)$$

which is equivalent to,

$$I = (I - A)(A^m + A^{m-1} + \ldots + A + I) \qquad (*)$$

and hence the inverse of $(I - A)$ is

$$(A^m + A^{m-1} + \ldots + A + I).$$

Check the right-hand side of the equation in (*) is in fact the identity, by expanding by hand or using DERIVE.

Chapter 2

Exercise 2A

Matrix of coefficients is $\begin{bmatrix} 2 & 3 \\ 3 & 5 \end{bmatrix} = A$ (say). Then DET $A = 2 \times 5 - 3 \times 3 = 1$. Hence the system has a unique solution. The solution is

$$x = \begin{vmatrix} 5 & 3 \\ 6 & 5 \end{vmatrix} = 7 \quad \text{and} \quad y = \begin{vmatrix} 2 & 5 \\ 3 & 6 \end{vmatrix} = -3$$

Exercise 2B

$$y = \frac{k_1 a_{31} a_{23} - k_1 a_{21} a_{33} + k_2 a_{11} a_{33} - k_2 a_{13} a_{31} + k_3 a_{13} a_{21} - k_3 a_{11} a_{23}}{a_{11} a_{22} a_{33} - a_{11} a_{23} a_{32} + a_{21} a_{13} a_{32} - a_{12} a_{21} a_{33} + a_{12} a_{23} a_{31} - a_{13} a_{22} a_{31}}$$

$$z = \frac{k_1 a_{21} a_{32} - k_1 a_{22} a_{31} + k_2 a_{12} a_{31} - k_2 a_{11} a_{32} + k_3 a_{11} a_{22} - k_3 a_{12} a_{21}}{a_{11} a_{22} a_{33} - a_{11} a_{23} a_{32} + a_{21} a_{13} a_{32} - a_{12} a_{21} a_{33} + a_{12} a_{23} a_{31} - a_{13} a_{22} a_{31}}$$

Exercise 2C

2. (i) $\begin{vmatrix} 1 & 0 & 0 \\ 2 & 3 & 6 \\ -1 & 4 & -1 \end{vmatrix} = -27$

(ii) $\begin{vmatrix} 0 & 1 & 2 \\ -1 & 2 & 0 \\ 3 & 4 & -2 \end{vmatrix} = -2 - 20 = -22$

(iii) $\begin{vmatrix} 5 & -1 & 3 \\ 2 & -2 & 1 \\ 4 & 0 & 7 \end{vmatrix} = -36$

Exercise 2D

$$\text{cof } a_{43} = 2$$

$$\text{cof } a_{22} = 12$$

$$\text{cof } a_{34} = -3$$

Exercise 2E

(i) $\text{cof } (a_{31}) = 0$

$\text{cof } (a_{32}) = -6$

$\text{cof } (a_{33}) = 3$

$\text{DET}(A) = -27$

(ii) $\text{cof } (a_{31}) = -4$

$\text{cof } (a_{32}) = -2$

$\text{cof } (a_{33}) = 1$

$\text{DET}(A) = -22$

(iii) $\text{cof } (a_{31}) = 5$

$\text{cof } (a_{32}) = 1$

$\text{cof } (a_{33}) = -8$

$\text{DET}(A) = -36$

Consolidation Exercise

1. (i) -9

(ii) 178

For (i) by hand, notice that when 1 is in the first column it can be used to zero other elements in that column, using result (v) in section 2.4

$$\begin{vmatrix} 1 & -1 & 2 \\ 2 & 1 & 1 \\ 1 & -3 & 1 \end{vmatrix}$$ becomes $$\begin{vmatrix} 1 & -1 & 2 \\ 0 & 3 & -3 \\ 0 & -2 & -1 \end{vmatrix}$$ Row 1 stays the same

Row 2 - 2 (Row 1)

Row 3 - Row 1

Now we expand the last determinant by the first column which amounts to working out the 2×2 determinant in the bottom right-hand corner,

$$(3 \times -1) - (-3 \times -2) = -9$$

For (ii) we can do the same thing as there is a 1 in position a_{21}. Here we have the following,

$$\begin{vmatrix} 2 & 3 & -5 \\ 1 & 7 & -2 \\ 5 & -11 & 2 \end{vmatrix}$$ becomes $$\begin{vmatrix} 0 & -11 & -1 \\ 1 & 7 & -2 \\ 0 & -46 & 12 \end{vmatrix}$$ Row 1 -2 (Row2)

Row 2 stays the same

Row 3 - 5 (Row 2)

Expanding by the first column again gives the following expansion,

$$-1 \begin{vmatrix} -11 & -1 \\ -46 & 12 \end{vmatrix} = -((-11 \times 12) - (-1 \times -46)) = -(-132 - 46) = 178$$

2. Observe here that the matrix of coefficients for the system in (i) is the matrix in 1(i) whose determinant we found to be -9. Hence there is a unique solution. Similarly the matrix of coefficients in (ii) is the same as the matrix in 1(ii) where the determinant was 178. By Cramer's rule determinants need to be evaluated.

Solutions

(i) $x = 1$, $y = 0$, $z = 0$. Once you know this is the solution its almost obvious by inspection!

(ii) $x = -41/178$, $y = -17/178$, $z = -169/178$.

3. Observe again that the 1s in the third column can be useful in reducing entries to zero.

$$\begin{vmatrix} a_1^2 & a_1 & 1 \\ a_2^2 & a_2 & 1 \\ a_3^2 & a_3 & 1 \end{vmatrix} \text{ becomes } \begin{vmatrix} a_1^2 & a_1 & 1 \\ a_2^2 - a_1^2 & a_2 - a_1 & 0 \\ a_3^2 - a_1^2 & a_3 - a_1 & 0 \end{vmatrix} \begin{matrix} \\ R2 - R1 \\ R3 - R1 \end{matrix}$$

Evaluating by the third column gives us the 2×2 determinant,

$$\begin{vmatrix} a_2^2 - a_1^2 & a_2 - a_1 \\ a_3^2 - a_1^2 & a_3 - a_1 \end{vmatrix}$$

We now observe that there are factors in both the first row and the second row. This will then give,

$$(a_2 - a_1)(a_3 - a_1) \begin{vmatrix} a_2 + a_1 & 1 \\ a_3 + a_1 & 1 \end{vmatrix}$$

which evaluates to,

$$(a_2 - a_1)(a_3 - a_1)((a_2 + a_1) - (a_3 + a_1)) = (a_2 - a_1)(a_3 - a_1)(a_2 - a_3)$$

This last expansion is equivalent to the answer when we remember that,

$$(a_2 - a_1) = -(a_1 - a_2)$$

Using DERIVE on this problem is quite interesting.
Input the matrix and author DET(#1).
If you ask DERIVE to Simplify or Factor (Trivial) it does not produce the result we are looking for.
If you ask DERIVE to Expand the determinant it produces all the brackets multiplied out.
However, if you ask DERIVE to Factor the determinant and then use the Squarefree option it produces the answer.
This shows the uses the factor option can be put to. It is suggested that the reader studies the manual for the full story.

4. (i)   −11
    (ii)   72
    (iii) 36

Some observations here could be useful in future.

(i)   Is immediate, $(1 \times 1) - (6 \times 2) = -11$.

(ii)   Using row three to zero the 3 and the 2 above it, reduces the work as in question 1.

(iii)   Observe that the matrix is upper triangular.

Expanding by the first column gives us simply 6 times a certain 3×3 determinant. The first column of our 3×3 contains a 1 and two zeros so expanding by the column gives us 1 times a certain 2×2 determinant which evaluates to 3×2. Hence we can in future, when we have a triangular matrix, evaluate its determinant by simply multiplying the elements on the leading diagonal.

In this question the answer is $6 \times 1 \times 3 \times 2 = 36$.

This is an example where evaluating by hand can actually be quicker than using DERIVE!

5.   (i)   0
       (ii)   –261
       (iii)   –6888

By hand a neat solution for (i) is by observing that if we take row two from row three we get a new row three, [3 3 1]. If we then subtract row one from row two we also get [3 3 1]. Hence we have two rows alike and that means we can obtain a row of zeros by one further subtraction. This then means the value of the determinant is zero.

For (ii) we changed just one entry, a_{33}. This stops the pattern as in (i) but we could do the same operations to get two zeros in the third row. This gives us the determinant,

$$\begin{vmatrix} 17 & 46 & 7 \\ 3 & 3 & 1 \\ 0 & 0 & 3 \end{vmatrix}$$

This evaluates to

$$3 \times \begin{vmatrix} 17 & 46 \\ 3 & 3 \end{vmatrix}$$

which reduces to

$$3 \times \begin{vmatrix} 17 & 29 \\ 3 & 0 \end{vmatrix}$$

by taking column one from column two.

Finally we have to work out $3 \times (0 - 3 \times 29) = -3 \times 87 = -261$.

For (iii) we can reduce the large values in column one by subtracting row two from row three and then subtracting row one from row two. This give us,

$$\begin{vmatrix} 1001 & 17 & 2 \\ 1 & 1 & 6 \\ 1 & 1 & -1 \end{vmatrix}$$

which reduces to

$$\begin{vmatrix} 1001 & 17 & 2 \\ 1 & 1 & 6 \\ 0 & 0 & -7 \end{vmatrix}$$

This evaluates by the third row, to give,

$$-7 \times (1001 - 17) = -7 \times 984 = -6888$$

6. This is similar to (3) if we use DERIVE.
    If we input the matrix and ask for the determinant we get the expression,

$$-a^{12} + 3a^8 - 3a^4 + 1$$

Factor squarefree gets the answer given.
By hand we can obtain a neat solution if we observe that for example

$$(\text{Row } 1 - a \times \text{Row } 2)$$

gives us three zeros in row one and what's more it gives us $(1 - a^4)$ in the top left-hand corner which we can take out as a factor, i.e.

$$(1-a^4)\begin{vmatrix} 1 & 0 & 0 & 0 \\ a^3 & 1 & a & a^2 \\ b & a^3 & 1 & a \\ c & d & a^3 & 1 \end{vmatrix}$$

Repeat this idea by taking 'a' lots of row three from row two to get two zeros in the second row and $(1 - a^4)$ on the diagonal.

Repeat once again taking 'a' lots of row four from row three and you should have the determinant,

$$(1-a^4)\begin{vmatrix} 1 & 0 & 0 & 0 \\ a^3-ab & 1-a^4 & 0 & 0 \\ b-ac & a^3-ad & 1-a^4 & 0 \\ c & d & a^3 & 1 \end{vmatrix}$$

The resulting matrix is lower triangular and therefore from our discussion above in 4(iii) we can evaluate this directly by multiplying the diagonal elements together. This gives us the result $(1 - a^4)^3$ as required.

7.   The result $\text{DET}(\alpha A) = \alpha^n \text{DET}(A)$ comes directly from factorizing the determinant of αA. Each row of $\text{DET}(\alpha A)$ has a factor of α and by the rule of factorizing a determinant it means that n, α's will be brought out of the determinant. Hence α^n is the total factor with $\text{DET}(A)$.

Chapter 3

Exercise 3A

$$|[1, \ 0, \ 1]| = \sqrt{2}$$

$$\left.\begin{array}{l} |[2, \ -1, \ 3]| = \sqrt{14} \\[2mm] |[1, \ 2, \ -3]| = \sqrt{14} \\[2mm] |[-2, \ 1, \ -3]| = \sqrt{14} \end{array}\right\} \quad \text{3 of them are equal in length}$$

$$|[3, \ -2, \ 0]| = \sqrt{13}$$

$$\left|[1, \quad 1, \quad 1]\right| = \sqrt{3}$$

$$\left|[3, \quad 0, \quad -3]\right| = \sqrt{18}$$

$$\left|[3, \quad 4, \quad 7]\right| = \sqrt{73}$$

Exercise 3B

$\mathbf{a} = (1, 1, 3), \mathbf{b} = (0, -3, 9)$
$\mathbf{a} + \mathbf{b} = (1, 1, 3) + (0, -3, 9) = (1, -2, 12)$

$\mathbf{a} - \mathbf{b} = (1, 1, 3) - (0, -3, 9)$ in coordinate form
$\quad\quad = (1, 4, -6)$

$\mathbf{a} - \mathbf{b} = [1\ 1\ 3] - [0\ -3\ 9]$ in matrix form
$\quad\quad = [1\ 4\ -6]$

$\mathbf{b} + \mathbf{a} = [1\ -2\ 12] = \mathbf{a} + \mathbf{b}$

Exercise 3D

2. $\quad \mathbf{a}{\cdot}\mathbf{b} = 5, \quad\quad \mathbf{a}{\cdot}\mathbf{c} = 3, \quad\quad \mathbf{a}{\cdot}\mathbf{d} = 2,$
    $\quad \mathbf{b}{\cdot}\mathbf{c} = 9, \quad\quad \mathbf{b}{\cdot}\mathbf{d} = 4, \quad\quad \mathbf{c}{\cdot}\mathbf{d} = 4.$

Exercise 3E

Any vector of the form $(-3\alpha, 2\alpha, \alpha)$ for any real α.

Exercise 3F

$\mathbf{a} \times \mathbf{b} = -4\mathbf{i} - 3\mathbf{j} + \mathbf{k}$
$\mathbf{b} \times \mathbf{c} = -2\mathbf{i} - 3\mathbf{j} + \mathbf{k}$
$\mathbf{d} \times \mathbf{e} = -3\mathbf{i} - \mathbf{j} + 2\mathbf{k}$
$\mathbf{a} \times \mathbf{e} = 0$
$\mathbf{b} \times \mathbf{a} = 4\mathbf{i} + 3\mathbf{j} - \mathbf{k}$
$\mathbf{c} \times \mathbf{d} = 2\mathbf{i} + 3\mathbf{j} - \mathbf{k}$

From above

$\mathbf{a} \times \mathbf{b} = -\mathbf{b} \times \mathbf{a}, \mathbf{b} \times \mathbf{c} = -\mathbf{c} \times \mathbf{b}$

$\mathbf{a} \times \mathbf{e} = 0 \Rightarrow \mathbf{a}$ and $\mathbf{e}$ are parallel since $\mathbf{a} \neq \mathbf{0}$ and $\mathbf{e} \neq \mathbf{0}$.

DERIVE Activity 3D

(B)   (i) $\mathbf{i} \times \mathbf{i} = \mathbf{0}$
      (ii) $\mathbf{j} \times \mathbf{j} = \mathbf{0}$
      (iii) $\mathbf{k} \times \mathbf{k} = \mathbf{0}$
      (iv) $\mathbf{v} \times \mathbf{u} = 15\mathbf{i} + 3\mathbf{j} - 7\mathbf{k}$

Hence $\mathbf{u} \times \mathbf{v} = -\mathbf{v} \times \mathbf{u}$.

(C) $(1, 0, 1) \cdot (-1, 2, 3) = 2$
     $(1, 0, 1) \cdot (4, -1, 2) = 6$
     $(1, 0, 1) \cdot (2, -4, -6) = -4$
     $(-1, 2, 3) \cdot (4, -1, 2) = 0$ orthogonal
     $(-1, 2, 3) \cdot (2, -4, -6) = -28$
     $(4, -1, 2) \cdot (2, -4, -6) = 0$ orthogonal

Note: $(2, -4, -6) = -2(-1, 2, 3)$.

Cross products:

$(1, 0, 1) \times (-1, 2, 3) = -2\mathbf{i} - 4\mathbf{j} + 2\mathbf{k}$
$(1, 0, 1) \times (4, -1, 2) = \mathbf{i} + 2\mathbf{j} - \mathbf{k}$
$(1, 0, 1) \times (2, -4, -6) = 4\mathbf{i} + 8\mathbf{j} - 4\mathbf{k}$
$(-1, 2, 3) \times (4, -1, 2) = 7\mathbf{i} + 14\mathbf{j} - 7\mathbf{k}$
$(-1, 2, 3) \times (2, -4, -6) = \mathbf{0}$
$(4, -1, 2) \times (2, -4, -6) = 14\mathbf{i} + 28\mathbf{j} - 14\mathbf{k}$

$(-1, 2, 3)$ and $(2, -4, -6)$ are parallel (see note above).

Exercise 3I

For $\mathbf{p}$; $\theta = 61.87°$
For $\mathbf{r}$; $\theta = 65.16°$
For $\mathbf{s}$; $\theta = 39.23°$
For $\mathbf{t}$; $\theta = 85.27°$

Exercise 3J

2   (iii) is not a linear combination of vectors.

Exercise 3K

(i)   Dependent, $2v_1 - v_2 = v_3$.
(ii)   Dependent, $v_4 = 3v_1 - v_2 + v_3$.
(iii) Independent.

DERIVE Activity 3E

A set including the zero vector is always dependent.

Consolidation Exercise

1.   (i)   You are asked to solve an equation of the form,

$$[1 \ 7 \ -4] = \alpha[1 \ -3 \ 2] + \beta[2 \ -1 \ 1]$$

Here $\alpha = -3$ and $\beta = 2$.

This means the vector $[1 \ 7 \ -4]$ is dependent on the two vectors **u** and **v**, and can be written as a linear combination, $-3u + 2v$.

(ii)   Here you have to solve for α and β, if possible, the equation

$$[1 \ k \ 5] = \alpha[1 \ -3 \ 2] + \beta[2 \ -1 \ 1]$$

This is equivalent to the system of equations,

$$1 = \alpha + 2\beta$$
$$k = -3\alpha - \beta$$
$$5 = 2\alpha + \beta$$

The neatest solution is to find α and β from the first and third equation by the method of simultaneous equations. Your solution should be,

$$\alpha = 3 \quad \text{and} \quad \beta = -1.$$

These values must satisfy the second equation involving k for consistency. Substituting gives the value of k as –8.

N.B. The question asked for 'values' and left it open whether there was more than one value. As we have seen the only value k can take is –8.

(iii)   This part of the question generalizes the ideas in (i) and (ii). We are asked to solve the equation below for a, b and c.

$$[a\ b\ c] = \alpha[1\ {-3}\ 2] + \beta[2\ {-1}\ 1]$$

This is equivalent to three equations as above.

$$a = \quad \alpha + 2\beta$$
$$b = -3\alpha - \ \beta$$
$$c = \quad 2\alpha + \ \beta$$

Adding the last two equations together gives,

$$b + c = -\alpha$$

Hence we have α in terms of a and b.

We now have to get β in terms of a, b and c if possible.
Staying with the last two equations we substitute the value of α into the last equation to obtain,

$$\beta = c - 2\alpha = c + 2b + 2c = 2b + 3c.$$

We now use the first equation to get the desired relationship between a, b and c. Notice we left equation one out of our argument until this point so that we did not get ourselves into a circular argument. Hence

$$a = \alpha + 2\beta = -b - c + 2(2b + 3c)$$

i.e. $a = 3b + 5c$ or $a - 3b - 5c = 0$.

This last equation is the condition on a, b and c which ensures any vector satisfying this condition can be expressed as a linear combination of the given **u** and **v**.

Notice that the vector in (i) satisfies this equation. Put $a = 1$, $b = 7$ and $c = -4$. Furthermore if we had known this condition from the start we could have found the value of k quickly in (ii) by substituting $a = 1$, $b = k$ and $c = 5$ in our condition to obtain,

$$1 - 3k - 25 = 0.$$

This again shows that there is just one value for k, namely -8.

2. Do not be put off by matrices being vectors in this question!

   We are asked to express E in terms of A, B and C. This means we have to solve an equation of the form,

   $$E = \alpha A + \beta B + \gamma C.$$

   This is a matrix equation which yields four equations by equating elements. The four equations you should be able to write down at once are,

$3 = \alpha + \beta + \gamma$	by equating the elements in position a_{11}.
$-1 = \alpha + \beta - \gamma$	by equating the elements in position a_{12}.
$1 = \quad -\beta$	by equating the elements in position a_{21}.
$-2 = -\alpha$	by equating the elements in position a_{22}.

   Hence $\alpha = 2$, $\beta = -1$ and therefore $\gamma = 2$ from the first or second equations. Check them both as they have to be consistent. So we have the result,

   $$E = 2A - B + 2C$$

   and thus E is a linear combination of A, B and C.

3. Here we are asked the same question as in question 2 but the E has changed.

   The equations we now obtain are,

   $$2 = \alpha + \beta + \gamma$$
   $$1 = \alpha + \beta - \gamma$$
   $$-1 = \quad -\beta$$
   $$-1 = -\alpha$$

   Here we have from the last two equations, $\alpha = 1$ and $\beta = 1$.
   From the first equation this would give $\gamma = 0$.
   However if we now use the second equation with $\alpha = 1$ and $\beta = 1$ to check consistency, we find γ would have to be 1.
   As we are solving these equations simultaneously we cannot have two values for γ and therefore we must conclude that E is not a linear combination of the three matrices given.

4. Here we have to use the definition given in the chapter for linearly independent vectors. This means we take a linear combination of the given vectors and equate it to the zero vector.

i.e. $\alpha[1\ 1\ 1] + \beta[0\ 1\ 1] + \gamma[0\ 1\ -1] = 0$

We now have to show that the only solution to this equation is when

$\alpha = \beta = \gamma = 0.$

The vector equation gives three equations if we equate by coordinates.

The first coordinates give directly,	$\alpha = 0$
The second coordinates give the equation	$\alpha + \beta + \gamma = 0$
The third coordinates give the equation	$\alpha + \beta - \gamma = 0$

With $\alpha = 0$ in the last two equations we have

$\beta + \gamma = 0$

and $\beta - \gamma = 0$

from which by addition $2\beta = 0$ and by subtraction $2\gamma = 0$. Hence $\beta = 0$ and $\gamma = 0$. Thus we have shown that the only solution to our original equation is

$\alpha = \beta = \gamma = 0.$

5.   Recall that $\mathbf{u \cdot v}$ is worked out as $\mathbf{uv'}$ which gives us a scalar (a number). We are using ' for transpose as usual.

(i) $\mathbf{u \cdot (v + w)} = \mathbf{u(v + w)'} = \mathbf{u(v' + w')}$ by property of transpose

$\qquad\qquad\quad = \mathbf{uv' + uw'}$ by property of matrix multiplication

$\qquad\qquad\quad = \mathbf{u \cdot v + u \cdot w}$ by definition of scalar product

(ii) $\alpha(\mathbf{u \cdot v}) = \alpha\mathbf{uv'} = (\alpha\mathbf{u})\mathbf{v'}$ by matrix property

$\qquad\qquad = (\alpha\mathbf{u}) \cdot \mathbf{v}$

Similarly

$\alpha(\mathbf{u \cdot v}) = \alpha\mathbf{uv'} = \mathbf{u}(\alpha\mathbf{v'})$ by matrix property

$\qquad\qquad = \mathbf{u} \cdot (\alpha\mathbf{v})$ by definition of scalar product.

6. In this question you are introduced to a definition which you might not have met before, and then you are asked to use it to prove something else. This is quite a common occurrence in mathematics. The solution is very short and relies on the definition of the scalar product in the form,

$$\mathbf{u} \cdot \mathbf{v} = |\mathbf{u}|\,|\mathbf{v}|\cos(\theta)$$

Replace $|\mathbf{u}|\cos(\theta)$ by $\mathbf{u} \cdot \mathbf{v}/|\mathbf{v}|$ in the given definition and collect the two $|\mathbf{v}|$'s together in the denominator.

For the final part recall that $\mathbf{v} \cdot \mathbf{v} = |\mathbf{v}|\,|\mathbf{v}|\cos(0) = |\mathbf{v}|^2$.

7. From the hint given in the question you should have the equation,

$$\mathbf{a} + \mathbf{b} + \mathbf{c} = \mathbf{0}$$

Hence

$$\mathbf{a} \times (\mathbf{a} + \mathbf{b} + \mathbf{c}) = \mathbf{a} \times \mathbf{0} = 0.$$

But

$$\mathbf{a} \times (\mathbf{a} + \mathbf{b} + \mathbf{c}) = \mathbf{a} \times \mathbf{a} + \mathbf{a} \times \mathbf{b} + \mathbf{a} \times \mathbf{c}$$

(given in the question).

Furthermore $\mathbf{a} \times \mathbf{a} = \mathbf{0}$ by definition of vector product. Hence putting all this information together we have,

$$0 = \mathbf{a} \times \mathbf{b} + \mathbf{a} \times \mathbf{c}.$$

Alternatively

$$\mathbf{a} \times \mathbf{b} = -\mathbf{a} \times \mathbf{c} = \mathbf{c} \times \mathbf{a}$$

(by definition of vector product).

Remember vector products are not commutative.

Now expanding the equation $\mathbf{a} \times \mathbf{b} = \mathbf{c} \times \mathbf{a}$ we have,

$$|\mathbf{a}|\,|\mathbf{b}|\sin(\gamma) = |\mathbf{c}|\,|\mathbf{a}|\sin(\beta)$$

Knowing that |a|, sin(γ) and sin(β) are non-zero we can write the above equation in the form,

$$\frac{|b|}{\sin(\beta)} = \frac{|c|}{\sin(\gamma)}$$

Similarly we could start with $b \times (a + b + c)$ and obtain

$$\frac{|a|}{\sin(\alpha)} = \frac{|c|}{\sin(\gamma)}$$

Hence the result, with the abuse of notation that |a| is written as 'a' and similarly for |b| and |c|.

Chapter 4

Exercise 4A

1. $A^2 = I$, therefore $A^3 = A$ and this means A is periodic of period 2.
      $B^2 = B$ so B is idempotent **or** of period 1.

2.   Assuming $a \neq 0$ and $b \neq 0$, $a^k = 1$ **and** $b^k = 1$.

3.   If $cd = 1$ then E would be periodic of period 2.
      If $cd = \pm 1$ then E could be periodic of period 4.
      E cannot have odd period but it can have even period.

DERIVE Activity 4B

(A)   The matrix A satisfies $A^3 = A$, therefore A is periodic of period 2. Thus A is **not** idempotent.

(B) B has period 3.

(C)   If we want $A^2 = A$ (period 1 or idempotent) we have from algebra, matrices of the form

$$\begin{pmatrix} a & 1-a \\ a & 1-a \end{pmatrix}$$

DERIVE Activity 4C

(B) $B^3 = 0$ so B is nilpotent of index 3.
$C^2 = 0$ so C is nilpotent of index 2.

DERIVE Activity 4D

(B) A is involutary, B is **not** involutary as $B^2 \neq I$.
C is involutary.

Exercise 4C

(i) We have shown that $(AB)^{-1} = B^{-1}A^{-1}$ and DET AB = DET A DET B so

$$(AB)^{-1} = \frac{1}{\text{DET}AB} \text{adj}\,(AB)$$

and

$$B^{-1}A^{-1} = \frac{1}{\text{DET}B} \text{adj}\,B \times \frac{1}{\text{DET}A} \text{adj}\,A$$

Hence adj (AB) = adj B adj A.

(ii) From Result 1

$$\textbf{adj}\,A(\textbf{adj}(\textbf{adj}(A))) = \textbf{diag}(|\textbf{adj}\,A|,...)$$
$$= |\textbf{adj}\,A|I_n$$
$$= |A|^{n-1}I_n \quad \text{from Result 2.}$$

Hence multiplying both sides by A we have

$$A\,\text{adj}A(\text{adj}(\text{adj}(A))) = |A|^{n-1}.A$$

But A adj $A = |A|I_n$ so LHS becomes

$$|A|I_n\,(\text{adj}(\text{adj}(A))) = |A|^{n-1}A \quad \text{(RHS)}$$

therefore adj(adj(A))) $= |A|^{n-2}A$ as required.

Exercise 4D

$$C^{-1} = \begin{bmatrix} 2 & 3 \\ -1 & -1 \end{bmatrix}$$

$$B^{-1} = \begin{bmatrix} -\frac{6}{7} & -\frac{1}{7} & \frac{5}{7} \\ \frac{2}{7} & \frac{5}{7} & -\frac{4}{7} \\ \frac{3}{7} & -\frac{3}{7} & \frac{1}{7} \end{bmatrix}$$

Exercise 4E

(i) $B \sim \begin{bmatrix} 1 & -1 & 2 & 1 \\ 0 & 3 & -1 & 4 \\ 2 & 1 & 0 & -3 \\ 0 & 1 & 2 & 1 \end{bmatrix}$ $C \sim \begin{bmatrix} 5 & -1 & 2 & 0 \\ 2 & -1 & 3 & -4 \\ 0 & 0 & 1 & 2 \\ 1 & 0 & -1 & 1 \end{bmatrix}$

(ii) $B \sim \begin{bmatrix} 0 & 1 & 2 & 1 \\ 0 & 3 & -1 & 4 \\ 2 & -2 & -6 & -6 \\ 1 & -1 & 2 & 1 \end{bmatrix}$ $C \sim \begin{bmatrix} 1 & 0 & -1 & 1 \\ 2 & -1 & 3 & -4 \\ -3 & 0 & 4 & -1 \\ 5 & -1 & 2 & 0 \end{bmatrix}$

Exercise 4F

$$A^{-1} = \begin{bmatrix} \frac{3}{5} & \frac{1}{5} \\ -\frac{2}{5} & \frac{1}{5} \end{bmatrix}$$

$$B^{-1} = \begin{bmatrix} \frac{5}{17} & -\frac{1}{17} & -\frac{2}{17} \\ -\frac{8}{17} & \frac{5}{17} & -\frac{7}{17} \\ \frac{7}{17} & \frac{2}{17} & \frac{4}{17} \end{bmatrix}$$

Exercise 4G

Given matrix is $E_1^{-1} E_2^{-1} E_3^{-1} U$ where

$$E_1 = \begin{bmatrix} 1 & 0 & 0 \\ -2 & 1 & 0 \\ 0 & 0 & 1 \end{bmatrix}, \quad E_2 = \begin{bmatrix} 1 & 0 & 0 \\ 0 & 1 & 0 \\ -3 & 0 & 1 \end{bmatrix}, \quad E_3 = \begin{bmatrix} 1 & 0 & 0 \\ 0 & 1 & 0 \\ 0 & -2 & 1 \end{bmatrix}$$

and

$$U = \begin{bmatrix} 1 & -1 & 4 \\ 0 & 2 & -7 \\ 0 & 0 & 4 \end{bmatrix}$$

DERIVE Activity 4G

(C)   Ranks are 3, 3, 2 respectively.

Consolidation Exercise

1.   Using DERIVE input the given matrix and obtain its determinant, simplified. DERIVE should give you,

$$(1 - t^2).$$

Now if $DET(A) = 0$ no inverse can exist, so in this case if

$$1 - t^2 = 0$$

then A will not have an inverse, i.e. if $t = \pm 1$.

The inverse can be obtained from DERIVE directly which can be written as,

$$\frac{1}{1 - t^2} \begin{vmatrix} 1 & -t & -t \\ -t & 1 & 1 \\ -t & t^2 & 1 \end{vmatrix}$$

This format verifies the formula

$$A^{-1} = \frac{1}{DET(A)} adj(A)$$

2.   For a matrix A to be idempotent $A^2 = A$.

We will assume this is the case in what follows.

Considering $(I - A)^2$ we have,

$$(I - A)^2 = (I - A)(I - A) = I - A - A + A^2 = I - 2A + A = I - A$$

Here we used the fact that $A^2 = A$ and the laws of matrix algebra to simplify the final expression.

Since $(I - A)^2 = (I - A)$ we conclude that $(I - A)$ is idempotent.

Starting again with $(I - 2A)^2$ we have the following,

$$(I - 2A)(I - 2A) = I - 2A - 2A + 4A^2 = I - 4A + 4A = I.$$

Thus we see that $(I - 2A)$ is not idempotent but because we have an expression like $BB^{-1} = I$ we can say that $(I - 2A)$ is its own inverse.

Start again with $(I - 3A)^2$ we have,

$$(I - 3A)^2 = (I - 3A)(I - 3A) = I - 3A - 3A + 9A^2 = I + 3A.$$

So here $(I - 3A)$ is not its own inverse neither is it idempotent. Generalizing we have,

$$(I - nA)^2 = (I - nA)(I - nA) = I - nA - nA + n^2A^2 = I + (n^2 - 2n)A.$$

We see that when $n^2 - 2n$ is zero, $(I - nA)$ is its own inverse.

$$(n^2 - 2n) = n(n - 2) = 0 \quad \text{implies} \quad n = 0 \quad \text{or} \quad n = 2$$

$n = 0$ gives us the trivial case of $I^2 = I$ and the only other case is the one shown above when $n = 2$.

Also if $(I - nA)$ is idempotent then $-(n^2 - 2n) = n$ which implies that

$$n^2 - n = 0.$$

This has solutions $n = 0$ or $n = 1$.

Again the first solution yields the trivial result and the second gives the case we started with.

3. The third line of the 'proof' is where the error occurs. Even if $A \neq 0$ we cannot conclude $(A - B) = 0$ because matrices exist with the following property.

If $C \neq 0$ and $D \neq 0$ then $CD = 0$ is possible.

For example in the 2×2 case,

$$C = \begin{bmatrix} 2 & 0 \\ 3 & 0 \end{bmatrix} \quad \text{and} \quad D = \begin{bmatrix} 0 & 0 \\ 4 & 5 \end{bmatrix}$$

are both non-zero matrices yet $CD = 0$.

4. The matrix

$$A = \begin{bmatrix} 1 & -1 & 0 \\ 0 & 0 & 1 \\ 1 & 1 & 0 \end{bmatrix}$$

The columns are orthogonal to each other by using the scalar product and showing for any two vectors the answer is zero. For example

$$[1 \ 0 \ 1] \cdot [0 \ 1 \ 0] = 1 \times 0 + 0 \times 1 + 1 \times 0 = 0.$$

Observe also that the rows of this matrix are also orthogonal to each other. Using DERIVE or by hand the product

$$AA' = \text{diag}[2 \ 1 \ 2],$$

and $A'A = \text{diag}[2 \ 2 \ 1].$

From these expressions we can read off *a*, *b* and *c* and we observe that

$$AA' \neq A'A$$

i.e. *A* and its transpose do not commute in this case.

The inverse of *A* can be produced by DERIVE and is,

$$\begin{bmatrix} 1/2 & 0 & 1/2 \\ -1/2 & 0 & 1/2 \\ 0 & 1 & 0 \end{bmatrix}$$

Notice that $DET(A) = -2$ in this case.

Given any three mutually orthogonal vectors in $\Re^3$ we would in fact get similar results to the above. We will help you to show this using DERIVE in the general 3×3 case.

Input the general matrix

$$A = \begin{bmatrix} a & b & c \\ d & e & f \\ g & h & i \end{bmatrix}$$

On line 2 input the transpose of *A*.
Now compute *A'A* and consider the elements not on the leading diagonal. You will have expressions such as

$$ab + de + gh, \quad ac + df + gi \quad \text{and} \quad bc + ef + hi.$$

Now consider *A* and remember that the columns are mutually orthogonal. This leads to the conclusion that the three expressions above are all zero. Hence

$$A'A = \text{diag}[a^2 + d^2 + g^2, b^2 + e^2 + h^2, c^2 + f^2 + i^2].$$

Notice that the elements in this matrix are the squares of the lengths of the column vectors in *A*.

Now work out AA' in DERIVE and study the elements there. This time you will get expressions such as,

$$ad + be + cf, \quad ag + bh + ci \quad \text{and} \quad dg + eh + fi$$

and these will be zero if the rows of A are mutually orthogonal.

You will then be left with the diagonal matrix,

$$\text{diag}[a^2 + b^2 + c^2, d^2 + e^2 + f^2, g^2 + h^2 + i^2].$$

These are the square of the lengths of the row vectors of A. The two diagonal matrices thus formed are not necessarily equal as in our numerical solution. Go back now to the numerical solution and show that your general result holds with $a = 1$, $b = -1$ and so on.

5.   From the fact that the given matrix is upper triangular we can compute the determinant by multiplying the elements on the leading diagonal together.

(i)   The determinant is 1. The adjugate matrix is obtained by doing 16 calculations of 3×3 matrices. Most of the calculations are by inspection, because some of the cofactors are triangular.

The inverse is

$$\begin{bmatrix} 1 & -1 & -1 & 0 \\ 0 & 1 & -1 & -1 \\ 0 & 0 & 1 & -1 \\ 0 & 0 & 0 & 1 \end{bmatrix}$$

(ii)   Let the given matrix be B and put $B = I - A$ in the expression in Chapter 1, Consolidation Exercise 1D, problem 5. Then $A = I - B$ and by direct computation $A^4 = 0$ so $m = 3$. Hence

$$B^{-1} = (I - A)^{-1} = I + A + A^2 + A^3$$

which again gives the result above for the inverse.

(iv)   We assume you probably did this part first!

Chapter 5

Exercise 5B

If $\lambda = -1$ we have one equation $x_1 + x_2 = 0$ with an infinity of solutions, $(\alpha, -\alpha)$.
If $\lambda = 3$ we have one equation $x_1 = x_2$ with an infinity of solutions, (α, α).

Exercise 5C

(i) $x = 1, y = 2, z = 3$.
(ii) $x = 0, y = -1, z = 1$.

Exercise 5D

2. $x = \dfrac{9}{10}, \quad y = -\dfrac{23}{20}, \quad z = -\dfrac{5}{4}$.

Exercise 5E

(i) $x = \dfrac{9}{10}, \quad y = -\dfrac{23}{20}, \quad z = -\dfrac{5}{4}$.

(ii) $x = -\dfrac{1}{4}, \quad y = -\dfrac{55}{4}, \quad z = -\dfrac{31}{4}$.

DERIVE Activity 5C

(C)   Exact solution is $x_1 = -4, x_2 = 60, x_3 = -180, x_4 = 140$.

Exercise 5G

(i)   Next error term approximately,

$$[2.66748 \times 10^{-9}, 1.65562 \times 10^{-5}, -5.4729 \times 10^{-6}]^T.$$

Next estimate of solution is approximately,

$$[5.73792, 4.27556, 3.09034],$$

which to 3 significant figures agrees with previous solution.

(ii) (a) Approximate solution [0.928793, –3.54584].

(b) Approximate solution [7.38451, 6.32103, 12.92620].

Consolidation Exercise

1. The tabular display with the first and second equations interchanged and the new first row divided by 2, is given below. Remember pivotal condensation needs to find the largest number in modulus, in the first column, and then we divide throughout by it.

–1	1	1/2	2	5/2
0	2	1	3	6
1	–1	1	1	2
	2	1	3	6
	3/2	3		9/2

At this stage we can interpret the results although strictly we should divide by 2 on the penultimate line.

From the last line we have, $(3/2)x_3 = 3$ and hence $x_3 = 2$.

From the penultimate line we have,

$$2x_2 + x_3 = 3.$$

This gives $x_2 = 1/2$.

Finally from the first equation we have

$$-x_1 + x_2 + (1/2)x_3 = 2.$$

This gives,

$$x_1 = x_2 + (1/2)x_3 - 2 = -1/2.$$

Hence the solution is,

$$x_1 = -1/2, \quad x_2 = 1/2, \quad x_3 = 2.$$

2. Here we are not asked to use pivotal condensation so we do not have to re-order the equations unless we want to. We do not need to use row-sum checks either. We are asked to show that this system is inconsistent.

The tableau in this case looks as follows,

$$
\begin{array}{cccc}
1 & -2 & 1 & 3 \\
-1 & 4 & 0 & -1 \\
2 & 0 & 4 & 12 \\
\hline
2 & 1 & 2 \\
4 & 2 & 6 \\
\hline
0 & 2
\end{array}
$$

This last line gives us the information we are looking for. It says that

$$0x_3 = 2.$$

This is an impossible equation and therefore the system does not have a solution.

3.   The solution is $x_1 = 1$, $x_2 = 2$, $x_3 = 3$.

(i)   Cramer's rule can be used as there is a unique solution since the determinant of the matrix of coefficients is -20, i.e. non-zero.

Three determinants now need to be evaluated with the constants column, $[5 \ 3 \ -1]'$, replacing the first column of the original matrix, for the numerator of x_1, the second column for the numerator of x_2 and the third column for the numerator of x_3. Thus you should obtain,

$$x_1 = -20/-20 = 1, \quad x_2 = -40/-20 = 2 \quad \text{and} \quad x_3 = -60/-20 = 3.$$

(ii)   You need to solve $A\mathbf{x} = [5 \ 3 \ -1]'$ for $\mathbf{x}$ which is,

$$\mathbf{x} = A^{-1}[5 \ 3 \ -1]'$$

This again leads to $\mathbf{x} = [1 \ 2 \ 3]$ from which the answers can be read off.

4.   The given matrix is of the form

$$
\begin{bmatrix}
x & y & z \\
* & * & *
\end{bmatrix} = A
$$

and it is of rank two which means that the second row is not a multiple of the first row hence we can not get a row of zeros in any row-reduction process.

Now consider the augmented matrix $[A|b]$. If we row reduce this matrix it will still have rank 2 because the worst that can happen is that if $\mathbf{b} = [b_1 \ b_2]$ then $b_2 = \alpha b_1$ and thus the augmented matrix will be of the form,

$$\begin{bmatrix} x & y & z & b_1 \\ x' & y' & z' & 0 \end{bmatrix}$$

where not all the elements x', y', z' are zero from the first part. Hence $\text{Rank}(A) = \text{Rank}(A|b)$ and therefore we have consistency and an infinity of solutions, appealing to the criterion given in this chapter.

5. (i)   Input the matrix of coefficients and the column of zeros into DERIVE, i.e. input the augmented matrix $[A|0]$ (declare a 3×4 matrix).
Row reduction should give you no zero rows and therefore you can conclude that the rank of A is 3 and the rank of $[A|0]$ is also rank 3. Thus we must have a unique solution to this system. As it is unique, and we know such a system has at least one solution, we conclude that the solution is the trivial one,

$$x_1 = x_2 = x_3 = 0.$$

(ii)   Input the matrix of coefficients and the column of zeros as in (i). Row reduce this matrix and show that a row of zeros is obtained and thus the rank of A is 2 and the rank of $[A|0]$ is also 2. Hence we can conclude that there are an infinity of solutions for this system.

We can calculate these solutions from the reduced matrix given by DERIVE. You should have,

$$\begin{bmatrix} 1 & 0 & 1 & 0 \\ 0 & 1 & 1 & 0 \\ 0 & 0 & 0 & 0 \\ 0 & 0 & 0 & 0 \end{bmatrix}$$

Interpreting the first and second rows we have the equations,

$$x_1 + x_3 = 0 \quad \text{and} \quad x_2 + x_3 = 0.$$

Hence if $x_1 = \alpha$, the solutions are of the form,

$$[\alpha \quad \alpha \quad -\alpha].$$

6.   First round off to 3 significant figures to obtain the system,

$$x_1 + \quad x_2 = 50$$
$$x_1 + 1.03x_2 = 20$$

Now subtract the top equation from the bottom to get

$$0.03x_2 = -30$$

which gives

$$x_2 = -1000.$$

Hence from the first equation

$$x_1 = -1050.$$

If the original equations are assumed exact we can get exact solutions for x_1 and x_2 as, 15650/13 and −15000/13 respectively, by the same method as above. These approximate to 1203.85 and −1153.85 to 2 decimal places.

We observe the ill-conditioning at this stage by the large difference between the exact answers and those obtained by rounding.

Now the percentage relative errors in x_1 and x_2 are given by the exact expressions,

$$\frac{(15650/13-1050)}{15650/13} \times 100 \quad \text{and} \quad \frac{(-15000/13-(-1000))}{-15000/13} \times 100$$

respectively.

These reduce to (4000/313)% and (40/3)% respectively.   These in turn approximate to, 12.77955 and 13.33333 which to the nearest whole number give 13% in each case.

Chapter 6

Exercise 6B

2.   (i) S is a subspace of V.
      (ii) S is a subspace of V.
      (iii) S is not a subspace of V since zero vector is **not** present in S.

Exercise 6C

(i)   No, since $[1 \ 0 \ 1]^T$ is mapped to $[2 \ 5 \ 14 \ 18]$ in $\Re^4$ **not** the zero vector.

(ii)   Null space $= \{[\alpha, -\alpha, \alpha]; \alpha \in \mathbf{R}\}$.

DERIVE Activity 6C

(B)   Same as (ii) Exercise 6C. Dimension is 1. Basis given above.
      For 5×6 matrix; Rank is 5, number of variables is 6.
      Hence 1 free variable (x_6). Dimension of null space is 1.
      Basis for it is

$$\{[-15696 \ 1476 \ -1390 \ -7551 \ -8736]\}$$

      by choosing $x_6 = 7253$.

Exercise 6D

1.   In matrices, T would be the $n \times n$ identity matrix where n is the dimension of the vector space V.

2. T would be represented by an $m \times n$ matrix, A, where n is the dimension of V and m is the dimension of W. The entries of T would have to be such that given any vector $\mathbf{v}$, $A\mathbf{v} = \mathbf{0}$.

3.   (iii) $\begin{bmatrix} -1 & 0 \\ 0 & 1 \end{bmatrix}$ (iv) $\begin{bmatrix} 0 & -1 \\ 1 & 0 \end{bmatrix}$

    (v) (a) $\begin{bmatrix} -1 & 0 \\ 0 & -1 \end{bmatrix}$ (b) $\begin{bmatrix} 0 & 1 \\ -1 & 0 \end{bmatrix}$ (c) $\begin{bmatrix} 0 & 1 \\ 1 & 0 \end{bmatrix}$

Consolidation Exercise

1.   Remember that if a mapping is a linear transformation we ought to be able to
      write down the corresponding matrix.

      (i)   We would have to write down a 2 × 2 matrix here since F goes from $\Re^2$
             to $\Re^2$. We can do this since

             $$\begin{bmatrix} 1 & 1 \\ 1 & 0 \end{bmatrix}\begin{bmatrix} x \\ y \end{bmatrix} = \begin{bmatrix} x+y \\ x \end{bmatrix}$$

             Alternatively we would have to show that,

             $$F(\alpha_1 v_1 + \alpha_2 v_2) = \alpha_1 F(v_1) + \alpha_2 F(v_2)$$

             which would be true but tedious to write out.

      (ii)   In this case we cannot write a 1 × 2 matrix which gives us the same as
             the mapping F. Remember the matrix must have real number entries
             and not variables. Knowing it isn't possible we should now look for a
             counter-example or an axiom which fails.

             Consider the axiom $F(\alpha v) = \alpha F(v)$.
             This fails because, $F(\alpha v) = F(\alpha(x,y)) = F((\alpha x,\alpha y)) = \alpha x \alpha y = \alpha^2 xy$
             whereas, $\alpha F(v) = \alpha F((x,y)) = \alpha xy$.

             However a counter-example would be to let $v_1 = (1,2)$ and $v_2 = (3, 5)$
             then $v_1 + v_2 = (4,7)$ and $F(v_1 + v_2) = 28$. But $F(v_1) + F(v_2) = 2 + 15 =$
             $17 \neq 28 = F(v_1 + v_2)$

      (iii) This is linear and we can write down a 1 × 3 matrix to prove it.

             $$\begin{bmatrix} 2 & -3 & 4 \end{bmatrix}\begin{bmatrix} x \\ y \\ z \end{bmatrix} = \begin{bmatrix} (2x-3y+4z) \end{bmatrix}$$

      (iv)   This is not linear as the 1 in the first coordinate of the image vector
             upsets linearity. Here the neatest solution is to appeal to the fact that if

a mapping is linear it must send the zero vector in the domain to the zero vector in the codomain. Here,

$$F((0,0)) = (0 + 1, 0, 0 + 0) = (1, 0, 0) \neq (0, 0, 0).$$

(v)   This is not linear because of the modulus function.
Here we show that the axiom $F(v_1 + v_2) = F(v_1) + F(v_2)$ fails.
Let $v_1 = (x_1, y_2)$ and $v_2 = (x_1, y_2)$

$$F(v_1 + v_2) = F((x_1 + x_2,\ y_1 + y_2, z_1 + z_2))$$
$$= (|x_1 + x_2|, 0)$$

but

$$F(v_1) + F(v_2) = (|x_1|, 0) + (|x_2|, 0) = (|x_1| + |x_2|, 0).$$

Since $|x_1 + x_2| \neq |x_1| + |x_2|$, then $F(v_1 + v_2) \neq F(v_1) + F(v_2)$.

Alternatively a counter-example could be the following.

Let $v = (1, 2, 3)$ and $\alpha = -3$, then $\alpha v = (-3, -6, -9)$ and $F(\alpha v) = (3, 0)$.
However $\alpha F(v) = -3(1, 0) = (-3, 0) \neq F(\alpha v)$.

2.   Both f and g are linear since we can write down a 3×2 and a 2×3 matrix respectively.

For f we have the matrix

$$\begin{bmatrix} 1 & 1 \\ 1 & 2 \\ 3 & 0 \end{bmatrix}$$

and for g, the matrix

$$\begin{bmatrix} 1 & 1 & 1 \\ 2 & 0 & 3 \end{bmatrix}$$

Alternatively prove the single axiom,

$$F(\alpha_1 v_1 + \alpha_2 v_2) = \alpha_1 F(v_1) + \alpha_2 F(v_2)$$

holds.

For the composition $(g \circ f)$ we have by definition,

$$(g \circ f)(x, y) = g(f(x, y)) = g(x + y, x + 2y, 3x)$$

$$= (5x + 3y, 11x + 2y)$$

The matrix which represents this mapping is 2×2 because the domain of $(g \circ f)$ is the domain of f i.e. $\mathfrak{R}^2$. Furthermore the codomain of $(g \circ f)$ is the codomain of g which is also $\mathfrak{R}^2$.

The matrix is,

$$\begin{bmatrix} 5 & 3 \\ 11 & 2 \end{bmatrix}$$

The connection between the three matrices is,

$$\begin{bmatrix} 1 & 1 & 1 \\ 2 & 0 & 3 \end{bmatrix} \begin{bmatrix} 1 & 1 \\ 1 & 2 \\ 3 & 0 \end{bmatrix} = \begin{bmatrix} 5 & 3 \\ 11 & 2 \end{bmatrix}$$

The matrices representing the linear mappings, M_g and M_f are multiplied together in the correct order to obtain the matrix for the composition, $M_{g \circ f}$ i.e. $M_g M_f = M_{g \circ f}$

In some books matrix multiplication is defined by considering composition of linear mappings.

Kernel and Images

(a)   Consider f and its kernel, $\{(x,y): f(x,y) = 0\}$

$$\text{Ker}(f) = \{(x, y): x + y = x + 2y = 3x = 0\}$$

Since $3x = 0$ we have $x = 0$ and thus $y = 0$ and therefore the kernel of f is trivial i.e. $\{(0, 0)\}$. Its dimension is zero.

The image of f is the space spanned by its columns, i.e. the column space of M_f. Since the columns are independent the image is spanned by,

$$\{(1, 1, 3), (1, 2, 0)\}$$

Thus the image is two-dimensional.

Since the domain of f is $\mathfrak{R}^2$ it has dimension 2. We verify the equation given, $2 = 0 + 2$, in this case.

(b)   Consider g and its kernel, $\{(x, y, z): g((x, y, z)) = 0\}$.

This is equivalent to the system of homogeneous equations,

$$x + y + \quad z = 0$$

$$2x \quad\quad + 3z = 0$$

This system reduces to,

$$x + \quad y + z = 0$$

$$- 2y + z = 0$$

This system has one degree of freedom, i.e. choose z to be any value then y and x can be calculated. Alternatively the system has two leading variables and one free variable. Choosing $z = 2$ means that y must be 1 which in turn means x must be –3. Hence

$$\text{Ker}(g) = \{(-3, 1, 2)\}$$

and its dimension is one.

The image of g is the column space of M_g which is the space spanned by,

$$\{(1, 2), (1, 0), (1, 3)\}$$

Because these vectors are in $\mathfrak{R}^2$ we know three vectors are two many for a basis so there must be dependency. The rank of M_g is 2 by inspecting the rows. (Remember: row rank = column rank). Hence the image of g is two-dimensional and we can choose any two of the three vectors as a basis. The domain of g is $\mathfrak{R}^3$ and thus three-dimensional. Again, we verify the formula given, $3 = 1 + 2$, in this case.

(c)   Consider $(g \circ f)$ and its kernel, $\{(x, y): (g \circ f)(x, y) = 0\}$.

This is equivalent to solving the homogeneous system of equations,

$$5x + 3y = 0$$

$$11x + 2y = 0$$

Because $\mathrm{DET}(M_{g \circ f}) \neq 0$ we know from our theory that this system can only have the trivial solution so the kernel again contains only the zero vector and is zero-dimensional. The image of $(g \circ f)$ is the spaced spanned by the columns of $M_{g \circ f}$ which is two-dimensional by inspection. (One column is not a multiple of the other.) The columns themselves form a basis for the image. We verify the given formula again, $2 = 0 + 2$ in this case.

3.   (i)   The dimension is 2 since the two vectors are independent, i.e. one is not a multiple of the other.

(ii)   The dimension is 1 since the second polynomial is twice the first.

(iii)   The dimension is 2 since one is not a multiple of the other.

(iv)   The dimension is 2 since one matrix is not a multiple of the other.

4.   (i)   This linear transformation has 1×3 matrix to represent it, namely

$$[2 \; -3 \; 5]$$

From this its image must be 1 dimensional as its column and row rank is 1. A basis for its column space is $\{1\}$. Its kernel is the solution to the equation

$$2x - 3y + 5z = 0.$$

This has one leading variable and two free variables, i.e. two degrees of freedom. Hence the kernel has dimension 2.

We find a basis as follows.

Choose $y = 0$ and $z = 2$ then $x = -5$.
Choose $z = 0$ and $y = 2$ then $x = 3$.

This gives us two vectors, $[-5 \ 0 \ 2]$ and $[3 \ 2 \ 0]$ which are independent and span the kernel.

Check that $F((-5, \ 0, \ 2)) = F((3, \ 2, \ 0)) = 0$. Also observe that the dimension theorem for linear transformations is again verified since,

Dimension of the domain = 3,

Dimension of Ker = Nullity = 2

and   Dimension of Image = Rank = 1.

(ii)   In this case a 2×3 matrix represents the transformation. Explicitly it is,

$$\begin{bmatrix} 1 & 1 & 0 \\ 0 & 1 & 1 \end{bmatrix}$$

The kernel is found by solving the system of equations,

$x + y = 0$
$y + z = 0$

These give $x = -y = z$ which implies one degree of freedom and therefore the dimension of the kernel is 1 and a basis for it could be $\{(1, -1, 1)\}$.

The image of this transformation has dimension 2 since the rows of the 2×3 matrix are independent. A basis for the image could be the standard basis $\{(1, 0), (0, 1)\}$, the first and last columns of our matrix.

5.   Considering the set U we observe that it is non-empty because it contains the zero vector $(0,0,0)$, apart from the two vectors given as examples. We need to show that U is closed under vector addition and scalar multiplication.

Addition

Consider two vectors (x, y, z) and (x', y', z') in U. This means, $x + y + z = 0$ and $x' + y' + z' = 0$. The sum of the two vectors is $(x + x', y + y', z + z')$.

Does this sum belong to U? It does if the sum of its coordinates equals zero.

Consider $x + x' + y + y' + z + z' = x + y + z + x' + y' + z' = 0$ since $x + y + z = 0$ and $x' + y' + z' = 0$.

Similarly we have to show that for any α (a scalar) and any vector v in U then αv is in U.

Let $v = (x, y, z)$ again, then $\alpha v = (\alpha x, \alpha y, \alpha z)$. Does this vector belong to U? Check it does by adding its coordinates and factorizing. Remember $x + y + z = 0$.

6. (i)   From the hint we see that the two given matrices have determinant zero, but the sum of the two matrices is the identity 2×2 matrix which has determinant 1. Hence U is not closed under vector addition, and thus cannot be a subspace of V.

    (ii) The hint says that the identity 2×2 matrix, I, is idempotent and thus every scalar multiple should be if U is a subspace of V.

    Consider $2I$, is it idempotent?

    $$(2I)^2 = 2I.2I = 4I \neq 2I.$$

    Thus U is not closed under scalar multiples and hence cannot be a subspace of V.

7. The easiest approach to this question is to consider the two matrices separately. Call the 3×4 matrix A and the 2×4 matrix B for ease of discussion.

    Using DERIVE row reduce both A and B and compare the results. The reduced matrix for A should look like the following,

$$\begin{bmatrix} 1 & 2 & 0 & 1/3 \\ 0 & 0 & 1 & -8/3 \\ 0 & 0 & 0 & 0 \end{bmatrix}$$

The reduce matrix for B is simply the first two rows of the above matrix. Since the non-zero rows of the reduced matrices for A and B are the same we can conclude that their row spaces are the same.

8.　The standard basis is $E = \{(1, 0, 0), (0, 1, 0), (0, 0, 1)\}$ and the basis in the question was labelled F.

(i)　The transition matrix from E to F, called P in the question is,

$$\begin{bmatrix} 1 & 1 & 1 \\ 1 & 1 & 0 \\ 1 & 0 & 0 \end{bmatrix}$$

because $(1, 1, 1) = 1(1, 0, 0) + 1(0, 0, 1) + 1(0, 0, 1)$ and similarly for the other two vectors in F. In other words the matrix is made up of the vectors given in F.

(ii)　The transition matrix from F to E called Q in the question is,

$$\begin{bmatrix} 0 & 0 & 1 \\ 0 & 1 & -1 \\ 1 & -1 & 0 \end{bmatrix}$$

since　$(1, 0, 0) = 0(1, 1, 1) + 0(1, 1, 0) + 1(1, 0, 0)$

and　$(0, 1, 0) = 0(1, 1, 1) + 1(1, 1, 0) - 1(1, 0, 0)$

and　$(0, 0, 1) = 1(1, 1, 1) - 1(1, 1, 0) + 0(1, 0, 0)$.

(iii)　Compute PQ directly using DERIVE or by hand to show that $PQ = I$ and thus P and Q are inverses of each other.

9. (i)   For a matrix to be similar to itself we have to find an invertible matrix
            P such that $P^{-1}AP = A$. Here $P = I$ the $n \times n$ identity matrix fills this
            role since I is its own inverse and $I^{-1}AI = A$.

    (ii)   You are given that A is similar to B, so there exists a P such that
            $P^{-1}AP = B$. We are required to show there exists a Q such that
            $Q^{-1}BQ = A$, i.e. B is similar to A.

            From the equation $P^{-1}AP = B$ we get $AP = PB$ and then $A = PBP^{-1}$.
            Comparing this equation with $Q^{-1}BQ = A$ we see that as long as we
            choose $Q = P^{-1}$ we have proved that B is similar to A.

    (iii)   Here we are given two pieces of information.

                    A is similar to B and B is similar to C.

            So there exist P and Q with the properties,

            $$P^{-1}AP = B \quad \text{and} \quad Q^{-1}BQ = C.$$

            We are required to show the existence of a matrix D such that,

            $$D^{-1}AD = C.$$

            We find such a D by combining the two given equations as follows,

            $$P^{-1}AP = B \quad \text{implies} \quad Q^{-1}P^{-1}APQ = Q^{-1}BQ = C.$$

            This give us the equation,

            $$Q^{-1}P^{-1}APQ = C.$$

            Comparing this with the equation $D^{-1}AD = C$, we see that we can
            choose

            $$D = PQ$$

            which exists and has an inverse, since $(PQ)^{-1} = Q^{-1}P^{-1}$.
            Hence we have proved (iii).

These three proofs show that similarity is an equivalence relation on a set of square matrices.

Chapter 7

Exercise 7A

1. $\lambda = \pm 1$ are the eigenvalues.
    For $\lambda = 1$, eigenvectors of the form $[-3\alpha, \alpha]$.
    For $\lambda = -1$, eigenvectors of the form $[\alpha, -\alpha]$.

2. (i) Characteristic equation is $\lambda^2 - 4\lambda - 21 = 0$.
         Eigenvalues are 7 and -3.

    (iii) For $\lambda = 7$ eigenvectors of the form $[\alpha, 3\alpha]$.
          For $\lambda = -3$ eigenvectors of the form $[\alpha, -2\alpha]$.

DERIVE Activity 7A

(C) Characteristic polynomial is $\lambda^3 - 6\lambda^2 + 3\lambda - 1 = 0$.
     λ_1 is approximately 5.48641, $\lambda_2 = 0.256791 - 0.34166i$ and
     $\lambda_3 = 0.256791 + 0.341066i$.
     Using DERIVE, eigenvectors can be found.

Exercise 7B

For A. Charpoly $w^2 - 9w + 14$. Hence

$$A^2 - 9A + 14I = 0$$

and $$A^{-1} = \frac{9I - A}{14} = \begin{bmatrix} \frac{2}{7} & -\frac{1}{7} \\ -\frac{3}{14} & \frac{5}{14} \end{bmatrix}.$$

For B. Charpoly $w^3 - 12w^2 + 44w - 48 = 0$. Hence

$$B^3 - 12B^2 + 44B - 48I = 0$$

and $$B^{-1} = \frac{B^2 - 12B + 44I}{48}$$

$$B^{-1} = \begin{bmatrix} \frac{1}{3} & -\frac{1}{6} & \frac{1}{6} \\ -\frac{1}{8} & \frac{3}{8} & -\frac{1}{8} \\ \frac{1}{24} & \frac{1}{24} & \frac{5}{24} \end{bmatrix}$$

For C. Charpoly $w^3 - 4w^2 + 3w = 0$.

> Notice $w = 0$ is an eigenvalue which means $Cx = \lambda x = \mathbf{0}$. Therefore C **doesn't** have an inverse.

Exercise 7C

(i)   Trace is $3 = 1 + 2$.
        DET is $2 = 1 \times 2$.
        Eigenvalues are 1 and 2.

(ii)   Trace is 2.
        DET is -19.
        Eigenvalues approximate to $1.48891, -3.32583, 3.83691$.

(iii)   Trace is 6.
        DET is 15.
        Eigenvalues approximate to $5.21708, 2.26430$ and $-0.740694 \pm 0.849208i$.

Consolidation Exercise

1.   By hand we have to evaluate the 3×3 determinant,

$$\begin{vmatrix} 1-\lambda & 2 & 2 \\ 2 & 1-\lambda & 2 \\ 2 & 2 & 1-\lambda \end{vmatrix}$$

Using DERIVE we can use the CHARPOLY function to obtain the cubic. The Cayley-Hamilton theorem states that a matrix satisfies its own characteristic polynomial.

Work out the expression, $(A^3 - 3A^2 - 9A - 5I)$ using DERIVE to show that it gives the 3×3 zero matrix. Hence the theorem is verified. Use the Manage, Substitute facility of DERIVE for ease.

We could find the eigenvalues using DERIVE by factorizing the cubic expression given in the question. We could on the other hand observe that if $\lambda = -1$ then the cubic is satisfied and therefore we can state,

$$\lambda^3 - 3\lambda^2 - 9\lambda - 5 = (\lambda + 1)(\lambda^2 - 4\lambda - 5).$$

The quadratic factor reduces to $(\lambda - 5)(\lambda + 1)$. Hence the original cubic factorizes completely to,

$$(\lambda - 5)(\lambda + 1)^2.$$

The eigenvalues are therefore 5 and -1 (twice).

To answer the last part of the question we can use the MANAGE, SUBSTITUTE, facility of DERIVE again and experiment.

Does the matrix A satisfy $(A + I)^2 = 0$?
Does the matrix A satisfy $(A - 5I)(A + I) = 0$?
You should find the answers are No and Yes respectively.
Hence A satisfies the second degree polynomial, $\lambda^2 - 4\lambda - 5 = 0$.

2. The fact that this mapping is linear can be proved using the two axioms or the one alternative axiom given in the chapter. Also it is linear because we can write down the associated 2×2 matrix, namely

$$\begin{bmatrix} 3 & 3 \\ 1 & 5 \end{bmatrix}$$

Using DERIVE we find the eigenvalues of this matrix to be 2 and 6. The associated exact_eigenvectors are $[3 \ -1]$ and $[1 \ 1]$ respectively. Hence a basis for the eigenspace associated with the eigenvalue 2 is

$$\{[3 \ -1]\}$$

and a basis for the eigenspace associated with the eigenvalue 6 is

$$\{[1 \ 1]\}.$$

Both are therefore one-dimensional.

3.   Here the associated matrix is,

$$\begin{bmatrix} 0 & 1 \\ 1 & 0 \end{bmatrix}$$

The eigenvalues are 1 and –1 and the associated eigenvectors are [1 1] and [1 –1]. Again these can be bases for the one-dimensional eigenspaces.

4.   Here the associated matrix is,

$$\begin{bmatrix} 0 & 1 \\ -1 & 0 \end{bmatrix}$$

The eigenvalues for this matrix are not real. DERIVE will produce the two complex eigenvalues, i and $-i$ with the associated eigenvectors [1 i] and [1 $-i$] respectively. If we were asked to give real eigenvalues we would say this matrix hasn't got any.

5.   The clue for diagonalization is to look at the eigenvalues and eigenvectors of the given matrix.

(i)   The eigenvalues of the given 2×2 matrix are 0, twice, since the characteristic polynomial is $\lambda^2 = 0$. Check these results using DERIVE and remembering you will need to load VECTOR.MTH as a utility file. The corresponding exact eigenvectors are of the form [a $a/2$]. The eigenspace is one-dimensional and therefore the matrix cannot be diagonalized.

(ii)   Here we have two distinct eigenvalues namely, 2 and –3.   The corresponding eigenvectors are [a $a/4$] and [a $-a$] respectively. Hence we have two independent eigenvectors and thus the given 2×2 matrix is diagonalizable. The matrix P is made up of the eigenvectors (in its columns) and the diagonal matrix has the eigenvalues down its diagonal.

Here we can take

$$P = \begin{bmatrix} 1 & 4 \\ -1 & 4 \end{bmatrix} \quad \text{and} \quad D = \begin{bmatrix} -3 & 0 \\ 0 & 2 \end{bmatrix}$$

Check that $P^{-1}AP = D$.

(iii)   The given 3×3 matrix is upper triangular and therefore the characteristic equation is immediately written down in factorized form as,

$$(1 - \lambda)(1 - \lambda)(2 - \lambda) = 0.$$

Remember the determinant of a triangular matrix is simply the product of the elements on the leading diagonal. Hence this matrix has three real eigenvalues, 1 (twice) and 2.

If we look for the eigenvectors of $\lambda = 1$ we find we have vectors of the form $[x \quad y \quad 0]$ and thus we can produce two independent vectors, in particular we have a basis for the eigenspace given by the standard basis vectors, $\{(1, 0, 0), (0, 1, 0)\}$.

For $\lambda = 2$ we have eigenvectors of the form $[0 \quad 2z \quad z]$ and choosing $z = 1$ we have the eigenvector $[0 \quad 2 \quad 1]$. Hence the 3×3 matrix can be diagonalized.

$$P = \begin{bmatrix} 1 & 0 & 0 \\ 0 & 1 & 2 \\ 0 & 0 & 1 \end{bmatrix} \quad \text{and} \quad D = \begin{bmatrix} 1 & 0 & 0 \\ 0 & 1 & 0 \\ 0 & 0 & 2 \end{bmatrix}$$

Again check that $P^{-1}AP = D$.

(iv)   Again we can write down the characteristic equation in factorized form since the given matrix is lower triangular.

The equation is $(1 - \lambda)^3 = 0$ and this gives $\lambda = 1$ (thrice).

If we seek the eigenvectors we find they are of the form $[0 \quad 0 \quad a]$ and the eigenspace is one-dimensional. Hence the given matrix can not be diagonalized.

Chapter 8

Exercise 8A

3. $\alpha = \pm \dfrac{1}{\sqrt{3}}$

DERIVE Activity 8A

(B)   DET is –24 for the first matrix.
        DET is +1 for the second and matrix is orthogonal.
        DET is +1 for the third matrix but is **not** orthogonal.

Exercise 8B

2. $\mathbf{u}_1 = (1, 0, 0) = \mathbf{v}_1$. (It is already unit). $\mathbf{u}_2 = (0, 0, 1)$ $\mathbf{u}_3 = (0, 1, 0)$.

We have produced the standard basis which is orthonormal.

Consolidation Exercise

1.   If $\mathbf{u}$ is orthogonal to $\mathbf{v}$ it follows that the scalar or dot product of $\mathbf{u}$ and $\mathbf{v}$, $<\mathbf{u},\mathbf{v}>$, defined on the vector space from which these vectors come must be zero. We assume we are in a vector space of the form $\mathfrak{R}^n$ and the scalar product is the usual one.

We are given that $<\mathbf{u},\mathbf{v}> = 0$. We are asked to prove that $<k\mathbf{u},\mathbf{v}>$ is also zero. This is simply proved by the fact that $<k\mathbf{u},\mathbf{v}> = k<\mathbf{u},\mathbf{v}> = k0 = \mathbf{0}$.

For the final part we are given $\mathbf{v}_1$ and $\mathbf{v}_2$ and are asked to find a vector $\mathbf{u}$ in $\mathfrak{R}^3$ such that $<\mathbf{u},\mathbf{v}_1> = <\mathbf{u},\mathbf{v}_2> = \mathbf{0}$.

Using coordinates and letting $\mathbf{u} = (x, y, z)$ in $\mathfrak{R}^3$ we have from the above equations, $<\mathbf{u},\mathbf{v}_1> = <\mathbf{u},\mathbf{v}_2> = \mathbf{0}$, the following simultaneous equations,

$$x + y + 2z = 0$$
$$y + 3z = 0.$$

This is a homogeneous system with two leading variables, x and y, and one free variable, z. Choose $z = 1$, then $y = -3$ and $x = 1$. A vector orthogonal to the given vectors is therefore $(1, -3, 1)$.

The question asks for a unit vector so we need to find the length of this vector and divide by that length. Its length is

$$(1^2 + (-3)^2 + 1^2)^{\frac{1}{2}} = \sqrt{11}$$

Thus a unit vector orthogonal to v_1 and v_2 is

$$\left(\frac{1}{\sqrt{11}}, \frac{-3}{\sqrt{11}}, \frac{1}{\sqrt{11}} \right)$$

2. In this question we are asked to build up an orthogonal matrix having been given the first row as

$$\frac{1}{\sqrt{3}} \quad \frac{1}{\sqrt{3}} \quad \frac{1}{\sqrt{3}}$$

Let the second row be

$$x \quad y \quad z$$

then because these two rows have to be orthogonal we have from scalar products,

$$\frac{x}{\sqrt{3}} + \frac{y}{\sqrt{3}} + \frac{x}{\sqrt{3}} = 0$$

which is the same as, $x + y + z = 0$.

Choose $x = 0$ and $y = 1$ as we have two degrees of freedom or two free variables, x and y. Then $z = -1$. Hence $(0, 1, -1)$ is a vector orthogonal to our given first row, but it is not a unit vector so normalize it as we did in Q1 above. Its length is $\sqrt{2}$ and thus the second row is,

$$0 \quad \frac{1}{\sqrt{2}} \quad \frac{-1}{\sqrt{2}}$$

Now let the third row be

$$x \quad y \quad z.$$

We now have two equations which must be satisfied namely,

$$x + y + z = 0$$

and

$$y - z = 0.$$

In the first equation we have cancelled the $\sqrt{3}$s and in the second we have cancelled the $\sqrt{2}$ s. Again we have one free variable, z so choose it to be 1. Then $y = z = 1$ and $x = -2$. Hence $(-2, 1, 1)$ is a vector orthogonal to the first two rows. Again we must normalize it and in this case we divide by $\sqrt{6}$, its length. Hence the final matrix is,

$$
\begin{bmatrix}
\frac{1}{\sqrt{3}} & \frac{1}{\sqrt{3}} & \frac{1}{\sqrt{3}} \\
0 & \frac{1}{\sqrt{2}} & \frac{-1}{\sqrt{2}} \\
\frac{-2}{\sqrt{6}} & \frac{1}{\sqrt{6}} & \frac{1}{\sqrt{6}}
\end{bmatrix}
$$

Note that the columns taken two at a time are orthogonal. For example column 2 and column 3 have for their scalar product,

$$\left(\frac{1}{\sqrt{3}}\right)\left(\frac{1}{\sqrt{3}}\right) + \left(\frac{1}{\sqrt{2}}\right)\left(\frac{-1}{\sqrt{2}}\right) + \left(\frac{1}{\sqrt{6}}\right)\left(\frac{1}{\sqrt{6}}\right)$$

which gives $1/3 - 1/2 + 1/6 = 0$.

Further the columns are unit vectors since for example column 3 has for its length,

$$\sqrt{\left(\frac{1}{\sqrt{3}}\right)^2 + \left(\frac{-1}{\sqrt{2}}\right)^2 + \left(\frac{1}{\sqrt{6}}\right)^2} = \sqrt{\frac{1}{3} + \frac{1}{2} + \frac{1}{6}} = \sqrt{1} = 1$$

You should have verified that the columns form an orthonormal set and the rows were formed to make an orthonormal set. The matrix we produced was not unique as we had choices to make as we built the matrix up.

Finally take the determinant of the matrix which in our case is 1. Another one might have given -1 but these are the only two answers as we saw in the chapter.

3. This question is to give you simple practice in using the Gram-Schmidt method of orthogonalization.

Let $v_1 = (1, 1)$ and $v_2 = (2, 3)$ in $\Re^2$ be the given basis.

We want to produce u_1 and u_2 from v_1 and v_2, such that $\{u_1, u_2\}$ is an orthonormal basis for $\Re^2$. From v_1 we form u_1 by normalizing it. Hence

$$u_1 = \left(\frac{1}{\sqrt{2}}, \frac{1}{\sqrt{2}} \right)$$

Now we use the formula, $\lambda u_2 = v_2 - <v_2,u_1>u_1$ where $<v_2,u_1>$ means the scalar product of v_2 and u_1. Hence,

$$\lambda u_2 = (2, 3) - \left(\frac{5}{\sqrt{2}} \right)\left(\frac{1}{\sqrt{2}}, \frac{1}{\sqrt{2}} \right)$$

$$= (2, 3) - \left(\frac{5}{2}, \frac{5}{2} \right)$$

$$= \left(-\frac{1}{2}, \frac{1}{2} \right).$$

Hence we normalize this to find $\lambda = \frac{1}{\sqrt{2}}$ and thus

$$u_2 = \sqrt{2}\left(-\frac{1}{2}, \frac{1}{2} \right) = \left(-\frac{1}{\sqrt{2}}, \frac{1}{\sqrt{2}} \right)$$

Thus the required basis is,

$$\left\{ \left(\frac{1}{\sqrt{2}}, \frac{1}{\sqrt{2}} \right)\left(-\frac{1}{\sqrt{2}}, \frac{1}{\sqrt{2}} \right) \right\}$$

Notice that these vectors have unit lengths and that their scalar product is zero, as required.

4. This is similar to Q3 but a little harder as the vectors are in $\Re^3$.

Here let the three given vectors be v_1, v_2 and v_3 respectively. We will produce u_1, u_2 and u_3 from the three formulas given below and taken from the chapter.

$$\mathbf{u}_1 = \frac{\mathbf{v}_1}{|\mathbf{v}_1|}$$

$$\lambda\mathbf{u}_2 = \mathbf{v}_2 - <\mathbf{v}_2,\mathbf{u}_1>\mathbf{u}_1$$

$$\mu\mathbf{u}_3 = \mathbf{v}_3 - <\mathbf{v}_3,\mathbf{u}_1>\mathbf{u}_1 - <\mathbf{v}_3,\mathbf{u}_2>\mathbf{u}_2$$

Again we use the notation $<\mathbf{u},\mathbf{v}>$ for the scalar product of $\mathbf{u}$ and $\mathbf{v}$. Hence

$$\mathbf{u}_1 = \left(\frac{1}{\sqrt{3}},\frac{1}{\sqrt{3}},\frac{1}{\sqrt{3}}\right)$$

and

$$\lambda\mathbf{u}_2 = (1,2,3) - \left(\frac{6}{\sqrt{3}}\right)\left(\frac{1}{\sqrt{3}},\frac{1}{\sqrt{3}},\frac{1}{\sqrt{3}}\right)$$

$$= (1,2,3) - (2,2,2)$$

$$= (-1,0,1) .$$

Thus,

$$\mathbf{u}_2 = \left(-\frac{1}{\sqrt{2}},0,\frac{1}{\sqrt{2}}\right)$$

Finally,

$$\mu\mathbf{u}_3 = (2,-1,1) - \left(\frac{2}{\sqrt{3}}\right)\left(\frac{1}{\sqrt{3}},\frac{1}{\sqrt{3}},\frac{1}{\sqrt{3}}\right) - \left(-\frac{1}{\sqrt{2}}\right)\left(-\frac{1}{\sqrt{2}},0,-\frac{1}{\sqrt{2}}\right)$$

$$= (2,-1,1) - \left(\frac{2}{3},\frac{2}{3},\frac{2}{3}\right) - \left(\frac{1}{2},0,-\frac{1}{2}\right)$$

$$= \left(\frac{5}{6},-\frac{5}{3},\frac{5}{6}\right).$$

Hence

$$\mu = \sqrt{\left(\left(\frac{5}{6}\right)^2 + \left(-\frac{5}{3}\right)^2 + \left(\frac{5}{6}\right)\right)} = \sqrt{\left(\frac{25}{6}\right)} = \frac{5}{\sqrt{6}}$$

Thus we have,

$$u_3 = \left(\frac{\sqrt{6}}{5}\right)\left(\frac{5}{6}, -\frac{5}{3}, \frac{5}{6}\right) = \left(\frac{1}{\sqrt{6}}, -\frac{2}{\sqrt{6}}, \frac{1}{\sqrt{6}}\right)$$

Thus the orthonormal basis for $\mathfrak{R}^3$ obtained from the three given vectors is given by,

$$\left\{\left(\frac{1}{\sqrt{3}}, \frac{1}{\sqrt{3}}, \frac{1}{\sqrt{3}}\right)\left(-\frac{1}{\sqrt{2}}, 0, \frac{1}{\sqrt{2}}\right)\left(\frac{1}{\sqrt{6}}, -\frac{2}{\sqrt{6}}, \frac{1}{\sqrt{6}}\right)\right\}$$

Again observe that these vectors are unit length and any two are orthogonal.

Chapter 9

Exercise 9A

1. 3×2 matrix $\begin{bmatrix} 0.2 & 0.1 \\ 0.3 & 0.3 \\ 0.5 & 0.6 \end{bmatrix}$

   2×2 matrix $\begin{bmatrix} 0.4 & 0.2 \\ 0.6 & 0.8 \end{bmatrix}$

2.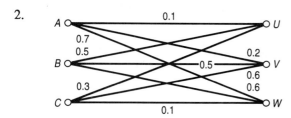

   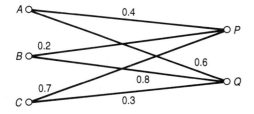

4. $A = \begin{bmatrix} a & 1-a \\ 1-a & a \end{bmatrix}$ $B = \begin{bmatrix} b & 1-b \\ 1-b & b \end{bmatrix}$ $\begin{array}{l} 0 \le a \le 1 \\ 0 \le b \le 1 \end{array}$

$\Rightarrow 0 \le 1-a \le 1$ and $0 \le 1-b \le 1.$

5. $A = \begin{pmatrix} 0.3 & 0.4 & 0.8 \\ 0.7 & 0.6 & 0.2 \end{pmatrix}$ and $B = \begin{pmatrix} 0.1 & 0.3 \\ 0.5 & 0.6 \\ 0.4 & 0.1 \end{pmatrix}$

11 arrives at P; 32.5 arrives at Q; 16.5 arrives at R.

Exercise 9B

	Coal	Steel	Lorries
TOTALS			
Coal 30 million tons	10	8	12
Steel 40 million tons	12	20	8
Lorries 20 million tons	6	4	10

Exercise 9C

1. $\begin{bmatrix} dx_1/dt \\ dx_2/dt \end{bmatrix} = \begin{bmatrix} 3 & -4 \\ -1 & -2 \end{bmatrix} \begin{bmatrix} x_1 \\ x_2 \end{bmatrix}$ $\begin{bmatrix} dx_1/dt \\ dx_2/dt \end{bmatrix} = \begin{bmatrix} 0 & 1 \\ 4 & 0 \end{bmatrix} \begin{bmatrix} x_1 \\ x_2 \end{bmatrix}$

2. $e^A = I + A + \dfrac{A^2}{2} + \dfrac{A^3}{6}$ (first 4 terms).

(i) $e^A = \begin{bmatrix} 2\frac{2}{3} & -3\frac{2}{3} \\ 0 & 6\frac{1}{3} \end{bmatrix}$

(ii) $e^A = \begin{bmatrix} 202\frac{2}{3} & 67 \\ 67\frac{2}{3} & 68\frac{2}{3} \end{bmatrix}$

Index of Definitions

We give below a quick reference list for some definitions which you might have to look up from time to time, until they are familiar. We advise you to make full use of the list of contents for the chapters as well as the sections within the chapters to look items up.

Bibliography for Further Reading and Sources for More Exercises

Below we give a selection of books, some old, some new, which contain theory as well as lots of exercises for you to work through using DERIVE. Some of the texts are included because they are classics in the subject.

As linear algebra is probably the area of mathematics which is most prolific for textbooks we advise you to look around in libraries for any book with linear algebra in the title. We are sure you will find some exercises and examples which are worthy of your attention.

Ayres, F. (1962). *Theory and Problems of Matrices*, Schaum's Outline Series, McGraw-Hill

Berry, J.S., Graham, E., Watkins, A.J.P. (1993). *Learning Mathematics Through DERIVE*, Ellis Horwood.

Birkhoff, G., MacLane, S. (1963). *A Survey of Modern Algebra*, Macmillan.

Blyth, T.S., Robertson, E.F. (1984). *Matrices and vector spaces*, Book 2 of Algebra through practice, Cambridge, Cambridge University Press.

Brickell, F. (1972). *Matrices and Vector Spaces*, No 8 in Problem Solver Series, George Allen and Unwin.

Dixon, C. (1971). *Linear Algebra*, Van Nostrand Reinhold.

Farkas, I., Farkas, M. (1975). *Introduction to Linear Algebra*, Adam Hilger.

Finkbeiner, D.T. (1966). *Introduction to Matrices and Linear Transformations*, 2nd edn, W.H. Freeman and Co.

Grossman, S.I. (1991). *Elementary Linear Algebra*, 4th edn, Saunders College Publishing, a division of Holt, Rinehart and Winston.

Halmos, P.R. (1958). *Finite-Dimensional Vector Spaces*, 2nd edn, Van Nostrand.

Hamilton, A.G. (1988). *A First Course in Linear Algebra with Concurrent Examples*, Cambridge, Cambridge University Press.

Hill, R.O. (1991). *Elementary Linear Algebra with Applications*, 2nd edn, Harcourt Brace

Johnson, L.W., Riess, R.D., Arnold, J.T. (1989). *Introduction to Linear Algebra*, 2nd edn, Addison-Wesley.

Kahn, P.J. (1967). *Introduction to Linear Algebra*, Harper and Row.

Lay, D.C. (1994). *Linear Algebra and Its Applications*, Addison-Wesley.

Lipschutz, S. (1968). *Theory and Problems of Linear Algebra*, Schaum's Outline Series, McGraw-Hill.

Neill, H., Moakes, A.J. (1967). *Vectors, Matrices and Linear Equations*, Oliver and Boyd.

Shilov, G.E. (1961). *An Introduction to the Theory of Linear Spaces*, Prentice Hall.

Towers, D. (1988). *Guide to Linear Algebra*, Macmillan Mathematical Guides, Macmillan Education.

Whitelaw, T.A. (1983). *An Introduction to Linear Algebra*, Blackie

Index